Maintenance organization and systems

To the maintenance managers of UK and Irish
industry, for their co-operation over the last twenty-five years
in providing much of the information that has allowed this
and my previous books to be written

Maintenance organization and systems

Dr Anthony Kelly

Butterworth-Heinemann
Linacre House, Jordan Hill, Oxford OX2 8DP
A division of Reed Educational and Professional Publishing Ltd

℞ A member of the Reed Elsevier plc group

OXFORD   BOSTON   JOHANNESBURG
MELBOURNE   NEW DELHI   SINGAPORE

First published 1997
© Reed Educational and Professional Publishing Ltd 1997

British Library Cataloguing in Publication Data
A catalogue record for this book is available from the British Library

ISBN 0 7506 3603 3

Library of Congress Cataloguing in Publication Data
A catalogue record for this book is available from the Library of Congress

Printed and bound in Great Britain by
Biddles Ltd, Guildford and King's Lynn

Contents

Preface

Business Centred Maintenance (BCM) is a structured approach to the overall task of managing the maintenance of physical assets, especially those of an industrial organization. BCM takes as its starting point the clear identification of the business aims. These aims are then translated into maintenance objectives which then form the basis of maintenance strategy formulation.

This is the second book of a two-volume presentation of the ideas of BCM. The first book, *Maintenance Strategy* (henceforth referred to as 'Book I') focused on the general methodology – on the setting of objectives and the formulation of the unit life plans and the plant maintenance schedule. Here, in Book II, we shall be concerned with the organizational, systems and documentational elements of the maintenance management task.

In Chapter 1 the overall methodology of BCM is developed via an industrial case study. In Chapter 2 it is then explained how a maintenance organization can be analysed into its three main elements: the resource structure, the administrative structure and the information and decision-making systems – and how it is influenced by external factors, such as human resource management policy, and internal ones, such as the maintenance workload (the forecasting, mapping and impact of which are dealt with in Chapter 3). Modelling of firstly the resource structure and then of the administrative structure is explained in Chapters 4 and 5 respectively, these chapters also identifying (a) the main factors (such as the use of contract labour) that influence the design of these structures, (b) the main problem areas (such as conflict with Production in the case of maintenance administration), and (c) some guidelines for improvement. Trends in maintenance organization that have arisen over the last twenty years – concerned with such ideas as self-empowerment, maintenance-production teams and decentralization – are reviewed in Chapter 6, the organizational section of the book then being rounded off by presentation, in Chapter 7, of various industrial case studies and exercises which reinforce the ideas covered up to this point.

The remainder of the book is concerned with maintenance management systems. Chapter 8 deals with the short-term planning and control of work (the key system in that, via it, much of the information is acquired which is needed for the operation of other systems, such as cost control). Chapter 9 deals with the managing of major plant shutdowns (showing how this is divisible into four main phases: preparation, planning, implementation and review, each phase being considered in some detail by referring to industrial examples). In Chapter 10 the three principal maintenance control systems are discussed, namely the control of cost and availability, of maintenance effectiveness and of organizational efficiency. Chapter 11 is a review of spare parts management and examines in particular the management of slow moving, high-cost items.

The final three chapters are concerned with maintenance documentation – Chapter 12 developing a model of it (via an explanation of the operation of the traditional paperwork system), Chapter 13 reviewing the basics of computer hardware and software and the final chapter detailing a procedure for selecting, implementing and commissioning a computerized maintenance system.

Acknowledgements

I am deeply indebted to my colleagues Peter Bulger, who most generously contributed Chapters 13 and 14, and John Harris, who edited the complete text (and was responsible for a major part of Chapter 11). I must also thank Tom Lenahan and Dr Harry Riddell, the former for revising and updating – based on his own extensive experience – the treatment of shutdown planning given in Chapter 9, and the latter for contributing Exercise 3 of Chapter 7.

The following have also contributed through discussion and correspondence arising out of my own and my IMMS partners' industrial consultancy work:

John Abbott and Brian Gover, Comalco, Australia
Alan Bonney, BHP Coal, Australia
Colin Bower, ex QEC, Australia
Glen Chuter, Alcan, Australia
Alan Dundass, Nabalco, Australia
Professor Richard Dwight, University of Wollangong, Australia
David Eiszele, John Collins, Bill Wallace, Western Power, Australia
Richard Grey, Courtaulds Chemicals, UK
Adrian Jones. Anglesey Aluminium
Peter Mackenzie, MINCOM, Australia
Jeff Miller, Peak Gold Mines, Australia
David McLatchie, Petroleum Refineries (Australia)
Tom Muldoon and Dermot Connellan, ESB, Ireland
Ray Parkin, Capcoal, Australia
Ian Roberts, ECNZ, New Zealand
Liam Tobin, Boyne Smelters, Australia
Barry Wilmer, Nissan, UK
Mark Zamitt, QAL, Australia

I would also like to thank:

- Bill Geraerds, Emeritus Professor of Industrial Engineering, University of Eindhoven, Holland, for help and advice which has greatly influenced my work;
- John Day, Alumax, USA, for informative discussions regarding his own approach to maintenance management;
- the many former students at the University of Manchester's School of Engineering whose research projects provided much of the information on which this book is based – in particular John Halstead, Greville Seddon, Chris Bull, Mohammed Al-Fouzan, Julia Gauntly and the various postgraduates seconded from the ESB, Republic of Ireland.

It has been my experience in presenting many industrial training and educational courses in maintenance management which prompted me to put this book together and I am therefore indebted to those who have helped in the organization of these, notably David Willson of Conference Communication, Dr Marwan Koukash of EuroMaTech, Michael Shiel of the Irish Management Institute, Len Bradshaw of EIT (Australia) and, in particular, Professor Erin Jancauskas, Dean of Engineering at Central Queensland University, Australia (for having the foresight and enthusiasm to set up the CQU postgraduate distance-learning degree in maintenance management).

Finally I have to thank Denise Jackson, Vicky Taylor, Carol Critchley and Beverley Knight, of the University of Manchester School of Engineering, for producing the figures and typescript, and especially for their remarkable forbearance when dealing with my countless alterations.

Anthony Kelly
IMMS
Bollin House
5 Edgeway
Wilmslow SK9 1NH
Cheshire UK
Tel 01625 529379
Fax 01625 539585

1
A business centred approach to maintenance management

Introduction

As explained in the preface, this is the second of two companion books on Business Centred Maintenance (BCM), which is the author's own preferred name for his particular approach to maintenance management – because it is an approach which is based throughout on the drive to achieve business objectives, which are translated into maintenance objectives and then used as the basis of strategy formulation. In the first book – *Maintenance Strategy* (henceforth referred to as 'Book I') – we dealt with the identification of objectives and the formulation of unit life plans; here we shall examine the organizational and systems aspects of maintenance management. Before doing so, however – and for the benefit of those who may not yet have read Book I – it will be useful to briefly recall the overall BCM methodology for developing a maintenance strategy (see Figure 1.1), a strategy which is based on well-established administrative management principles (see Figure 1.2).

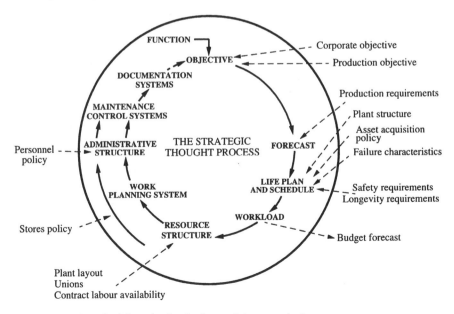

Figure 1.1   A methodology for developing maintenance strategy

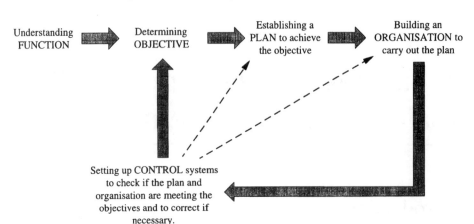

Figure 1.2   The basic steps of the management process

One way of describing the *function* of a maintenance department is to say that it is *ensuring and controlling the reliability of the plant*. The ways in which this function might be affected by its dynamic relationship with the production system need to be clearly understood. Once this has been achieved a definition of the maintenance *objective* that is compatible with the corporate and production objectives can be identified. This might well be, for example, as follows:

> to achieve the agreed plant operating pattern and product quality, within the accepted plant condition and safety standards, and at minimum resource cost

Consider, for example, the food processing plant (FPP) outlined in Figures 1.3 and 1.4, which show the layout and the process flow. The plant operates for fifteen continuous shifts per week, Monday to Friday. It is the responsibility of the FPP users to specify the product mix and output (in cans/week) they desire, and hence the maximum allowable downtime – and also various quality, safety and plant longevity requirements. The maintenance department is responsible for ensuring that – at minimum resource cost (i.e. of labour, materials, tools) – the plant is capable of meeting these requirements. The maintenance objectives need to be interpreted in a form that is meaningful at the main equipment level (that of a mixer, say, in the case of the FPP). This allows the maintenance *life plan* for the various units of plant to be established – on the basis of which the maintenance *schedule* for the plant is negotiated with production, taking into consideration the way in which the plant is used (the production policy often drives the maintenance schedule) and the level of plant redundancy. The main decision regarding the life plan is the determination of the nature and intensity of the preventive work – which, in turn, determines the resulting level of corrective work. The philosophy on which the FPP's maintenance life plan and schedule is based is outlined in Table 1.1.

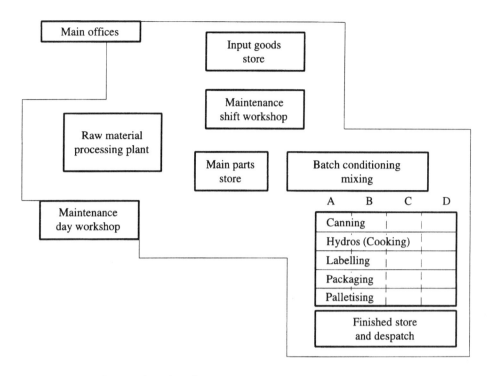

Figure 1.3   Food processing plant, layout

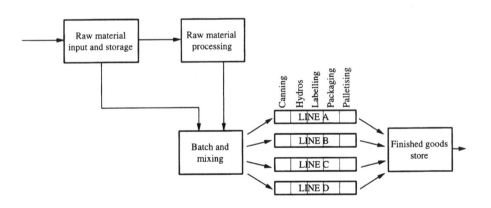

Pattern of operation: 50 weeks x 5 days x 3 shifts, Monday/Friday

Figure 1.4   Food processing plant, flow

Table 1.1 Food processing plant, maintenance philosophy

	Maintenance philosophy	Work type
Monday to Friday	"Keep the plant going" and "Keep an eye on its condition"	**Reactive maintenance** Operator monitoring Tradeforce line-patrolling Condition-based routines
Weekends	"Inspect the plant carefully and repair as necessary in order to keep it going until next weekend"	**Known corrective jobs** Inspect and repair jobs Fixed-time jobs
Summer shutdown	"Carry out the major jobs to see us through another year"	**Known corrective jobs** **Fixed-time major jobs**

The maintenance life plan and schedule strongly influences the level and nature of the *workload* (of preventive, corrective and modification work). The latter can be mapped by its organizational characteristics, i.e. its scheduling lead time, and for the FPP this is as shown in Figure 1.5. It can be seen that the workload – when considered alongside such factors as plant

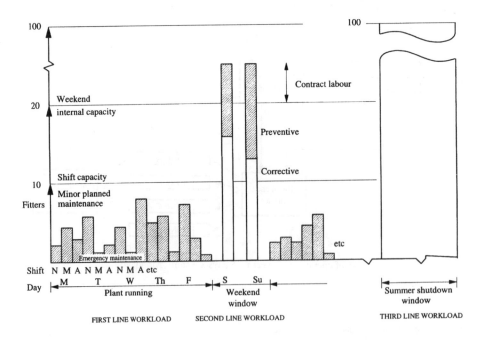

Figure 1.5   Workload pattern for fitters

layout – has a considerable influence on the nature and design of the maintenance organization.

The primary task of the maintenance organization is to match resources to workload, and in so doing to ensure that the agreed plant output is achieved at minimum resource cost – which is a restatement of the maintenance objective. In order to achieve this the organizational design needs to be aimed at maximizing tradeforce performance – which itself is a function of tradeforce utilization and motivation, of the availability of spares, tools and information and of work planning efficiency. Many inter-related decisions have to be made (Where to locate the manpower? How to extend inter-trade flexibility? Who should be responsible for spare parts? Who should be responsible for maintenance information?), each influenced by various conflicting factors. Thinking in terms of the framework of Figure 1.1 reduces the complexity of this problem, by categorizing the decisions according to the main elements of the organization, namely its resource structure, administrative structure, work planning system and so on, and then considering each element in the order indicated. The procedure is iterative.

The *resource structure* is the geographical location of workforce, spares, tools and information, their function, composition, size and logistics. For the FPP, for example, Figure 1.6 shows the Monday-to-Friday structure that has evolved to best suit the characteristics of a 24-hour first-line reactive workload. The emphasis is on rapid response, plant knowledge via specialization, shift working, and team-working with Production. The shift

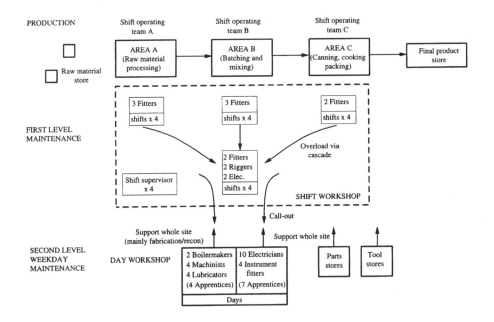

Figure 1.6 Weekday resource structure, food plant

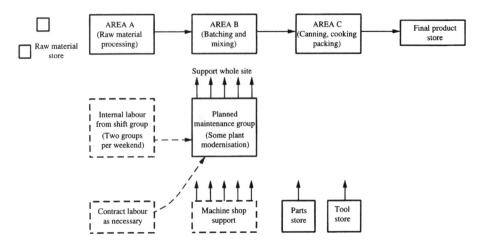

(Same structure for third line shutdown resource)

Figure 1.7   Weekend resource structure, food plant

groups have been sized to match the reactive workload. Lower priority jobs are used to smooth the workload.

Figure 1.7 shows the structure that matches the second-line weekend workload. This work is made up of relatively small preventive and corrective jobs that benefit from planning and scheduling. Contract labour is used to top-up, as necessary, the internal labour force. A similar approach is used for the annual shutdown, but in that case the contracted workforce exceeds that of internal labour.

The example shows the influence of the workload on the form of the organization. In practice, the purpose of any resource-structure design or modification will need to be specified and will most likely be expressed as being *to achieve the best resource utilization for a desired speed of response and quality of work*. In general, this comes down to determining the appropriate shape and size of the various trade groups (first, second and third line).

The second element in the design of a maintenance organization is the formation of a decision-making structure – the maintenance *administrative structure*. This can be considered as a hierarchy of work roles, ranked by authority and responsibility, for deciding what, when and how maintenance work should be carried out. An example is shown in Figure 1.8 (which uses the so-called organization chart as the modelling vehicle; many of the rules and guidelines of classical administrative theory can be used in the design of such structures). Here, the key decisions fall under two headings, lower and upper structure. Lower structure decisions are concerned mainly with establishing the work roles of the supervisors and planners and the relationship between these people and the production supervisors. The lower structure is initially considered separately from the upper because it is influenced by – indeed, almost constrained by – the nature of the maintenance resource structure

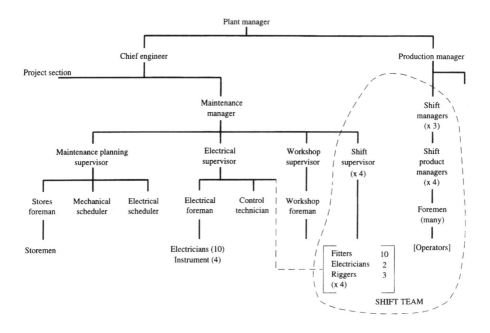

Figure 1.8   Administrative structure, food plant

which, as explained, is in turn a function of the workload.

The main task in establishing the upper structure is deciding on the relationship between the production group (the plant users), the maintenance supervisors and the trade groups (the maintenance doers), the maintenance planning group and the maintenance engineers (the technical decision makers). This complex problem is influenced by many factors, the most important of which are:

- the plant size, structure and geographical layout;
- the supervisory structure;
- the production administrative structure.

The third element in the design of a maintenance organization is the formation of the maintenance systems, the most important of which is the *work planning system*, which defines the way in which the work is planned, scheduled, allocated and controlled.

Figure 1.9 outlines such a system for the resource and administrative structure previously shown. The design of this should aim to strike the right balance between the cost of planning the resources and the savings in the direct and indirect maintenance costs that result from use of such resources.

A *control system* is needed to ensure that the maintenance organization is achieving its objectives and to provide corrective action, e.g. change the

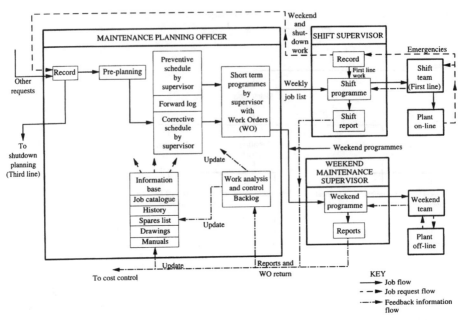

Figure 1.9   Work planning system, food plant

life plan, if it is not. There would appear to be three principal vehicles for this:

(i)   Control of maintenance *productivity*: ensuring that the budgeted levels of maintenance effort are being sustained and that required plant output is achieved.

(ii)   Control of maintenance *effectiveness*: ensuring that the expected long-term and short-term plant reliability is being achieved, i.e. that the life plan is effective and is being carried out.

(iii)   Control of maintenance *organizational efficiency*: monitoring the efficiency of utilization of workforce, materials and tools.

At this point it must be emphasized that the purpose of the organizational analysis that has been outlined is to enable the many decisions that affect the final shape of the organization to be seen in perspective. The possible effects of any particular decision on the complete organization and its objectives can then be better assessed. In the final analysis, the maintenance organization needs to be considered as a synergistic whole, i.e. an organism which is much more than merely the sum of its parts.

Figure 1.1 indicated that some form of formal *documentation* system – for the collection, storage, interrogation, analysis, and reporting of information (schedules, manuals, drawings or computer files) – is needed to facilitate the operation of all the elements of maintenance management. Figure 1.10, a general functional model of such a system (whether manual or computerized), indicates that it can be seen as comprising seven principal interrelated modules (performing different documentation functions). Considerable clerical and engineering effort is needed to establish and maintain certain of these functions (e.g. the preventive maintenance information base). The control module, in particular, relies on an effective data collection system.

An analytical framework, based on various system models and which can facilitate understanding of a maintenance system – proposed or existing – has been outlined. Any maintenance organization will, however, be designed and operated by people and will have many interfaces with other parts of the company (notably with Production) which are also made up of people. It is therefore essential that this systems structure approach is complemented by a full consideration of the *human factors* in an organization.

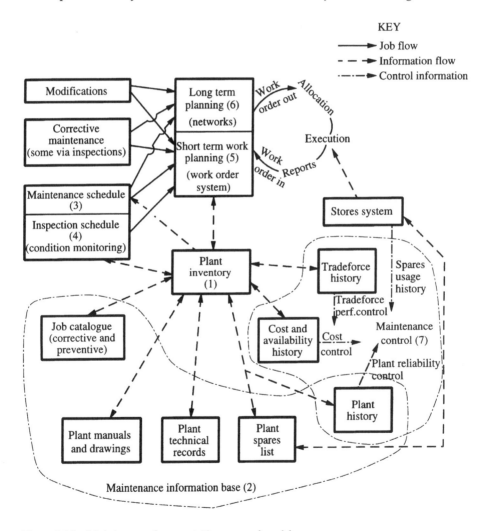

Figure 1.10   Maintenance documentation, general model

Application of the approach

The organizational mapping that has been described was carried out in order to advise the particular food processing company about changing their maintenance strategy to meet a changing production demand. At the time of the first visit the plant's production pattern was three shifts per day, five days per week, fifty weeks per year. There was also considerable spare capacity; for example, only three lines out of four (see Figure 1.4) were needed to achieve full capacity. However, each line had its own product mix to satisfy the market demand. Thus, the availability of any given line for maintenance depended on the market demand and the level of finished product stored (which could be up to two weeks output). Off-line maintenance could therefore be carried out in the weekend windows of opportunity or, by exploiting spare capacity, during the week. In general, most of the off-line work was carried out during the weekend.

The problem the company faced was that they were going to increase capacity to 21 shifts per week. They wanted to know how this was going to affect their maintenance department. The author was asked to map their existing maintenance management systems and propose an alternative strategy which would facilitate 21-shift operation.

The effect of 21-shift operation on maintenance

The existing maintenance strategy was based on carrying out off-line maintenance during the weekend windows and during the once-per-year holiday window. Little attempt had been made to exploit the excess capacity of the plant, or spare plant, to schedule off-line work while the plant was operating. The new 21-shift operating pattern meant that off-line maintenance would have to be carried out in this way. Indeed the strategy would have to move in the direction indicated in Table 1.2. This, in turn, would change the workload pattern, i.e.

- the first line workload would extent to 21 shifts,
- the off-line work (schedulable corrective and preventive) would need to be done during the weekday day-shifts,
- the third line major work could still be carried out in the holiday window.

Thus, to cover this workload, the maintenance organization would also have to change. The most likely organization would be based on a first-line, 21-shift group (perhaps with a reduction in manning-per-shift) and a second-line day-group operating five days per week. This, in turn, would influence the administrative structure and work planning systems. The latter would have to improve considerably.

The strategic thought process

The example shows that the maintenance department requires managerial strategic analysis in the same way as any other industrial department. The

Table 1.2 Changes in maintenance strategy to accommodate new production policy

(a) A movement towards shutdowns of complete sections of plant based on the longest running time of critical units (eg the Hydros). The frequency of these shutdowns will, as far as possible be based on running hours or cumulative output. However, for critical items, inspection and condition monitoring routines may be used to indicate the need for shutdowns, which will provide more flexibility and certainty about shutdown dates.

(b) All plant designated as non-critical, eg as a result of spare capacity, will continue to be scheduled at unit level (eg D line Filler/Closer).

(c) A much greater dependence on formalized inspections and condition monitoring routines, for reasons given in (a) and also to detect faults while they are still minor and before they become critical.

(d) A concerted effort either to design-out critical items or to extend their effective running time.

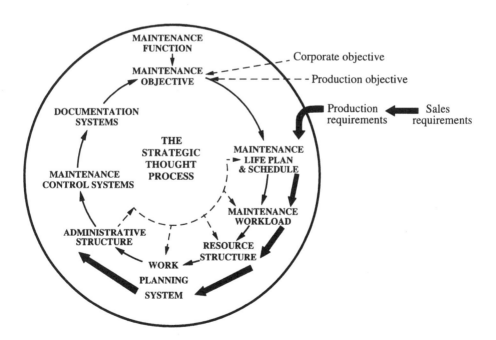

Figure 1.11 The influence of market demand on maintenance strategy

thought process that was involved is indicated in Figure 1.11. The process starts with the Sales–Production reaction to market demand, the resulting change in the plant operating pattern and the increased plant operation time. This, in turn, requires amended maintenance life plans and a modified maintenance schedule. Thus, the maintenance workload changes, which brings in train the need to modify the maintenance organization and systems. Understanding and applying this type of strategic thought process is the cornerstone of effective and fruitful maintenance management analysis.

2
Maintenance organization in outline

Introduction

The primary task of the maintenance organization is to match resources (men, spares, tools and information) to workload, so that the maintenance objective – sustaining, at minimum total cost, plant which is capable of producing the desired level and quality of output – can be attained. In order to achieve this the organization needs to be designed so that the performance of the tradeforce (a function of its utilization and motivation; of the availability of spares, tools and information; and of the efficiency of work planning) is maximized. Designing a maintenance organization therefore involves many interrelated decisions (where to locate the manpower; how to extend inter-trade flexibility; to whom to allocate responsibility for maintenance information, or for spare parts), each such decision being influenced by many conflicting factors.

The approach reviewed in Chapter 1 and outlined in simplified form in Figure 2.1 reduces the complexity of maintenance organizational design by categorizing the decisions according to the following main elements of the organization.

Structure	*The resource structure*: the location, mix, size, function and logistics of the maintenance resources – primarily the manpower (see, for example Figure 1.6).
	The administrative structure (the so-called organizational chart): the allocation of managerial responsibilities and interrelationships (see, for example, Figure 1.8).
Systems	*The short- and long-term work planning and work control system*: the maintenance costing system, etc. (see, for example, Figure 1.9).

Modelling the organization

In Chapter 1 it was shown, by considering the maintenance of a food processing plant, how the elements of a maintenance organization may be modelled. While there are clear benefits from undertaking such an analysis it is also important to understand how these elements interrelate to allow the organization to function. The *organizational whole* is greater than the sum of its *elemental parts* – it has *synergy*. One way of visualizing it is as a three-dimensional structure, as a *pyramid of personnel*.

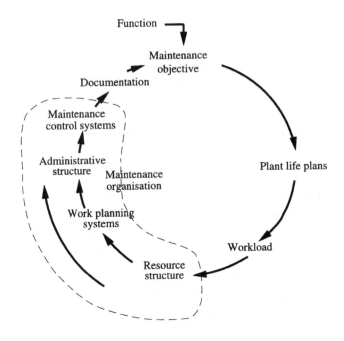

Figure 2.1   The main elements of maintenance organization

The maintenance tradesmen and the plant operators are at the base of the pyramid – *the resource structure* – and the management make up its remainder – *the administrative structure*. All the positions in the structure have work roles, i.e. duties, responsibilities, interrelationships, etc. (see Figures 2.2 and 2.3). The *work planning system* can be represented as an information and decision-making system running across the structure (see Figure 2.4). Other systems can be represented in a similar fashion.

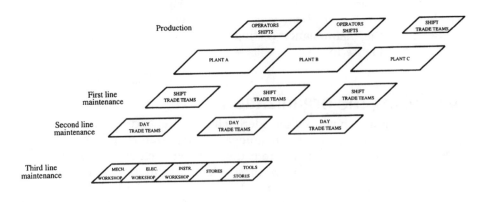

Figure 2.2   Three-dimensional model of an organization – the resource structure

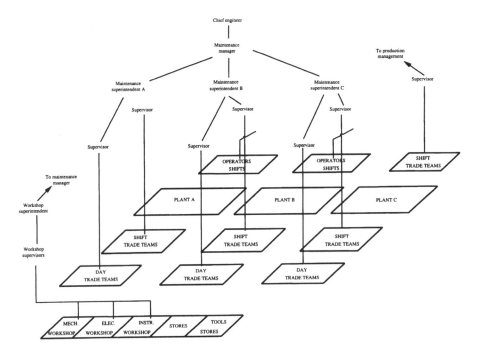

Figure 2.3   Three-dimensional model of an organization – the administrative structure

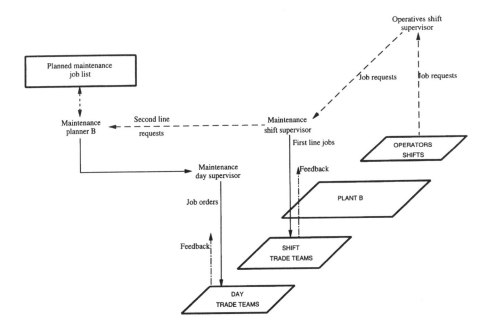

Figure 2.4   Work planning as a 'horizontal' information system

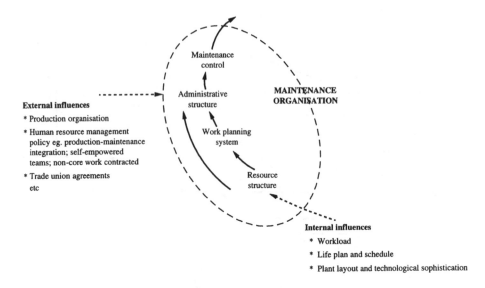

Figure 2.5   Influences on maintenance organizational design

Thus, when designing or modifying a maintenance organization the approach of Figure 2.1 needs to be followed, e.g. it has to be understood that the workload has a major influence on the resource structure which in turn influences the administrative structure and systems. These internal (bottom-up) influences on organizational design are often accompanied by external (top-down) influences, e.g. human resource management policy, concerning such matters as production-maintenance integration, the adoption of self-empowered teams, etc. (see Figure 2.5).

The final introductory point is that the organization must be 'dynamic', it needs to be responsive to environmental changes (both internal and external). Such changes can be revolutionary or, as is more often the case, evolutionary.

3
The maintenance workload

Introduction

The nature of the maintenance workload greatly influences the structure of the maintenance organization (see Figure 2.5 which delineates this relationship). Before considering the problem of forecasting and mapping this workload it will be instructive to outline some of its characteristics.

Workload characteristics

It has long been the custom to categorize maintenance work as being either preventive, corrective or modification. The last, although strictly not maintenance, is usually included because the maintenance department is often involved in carrying it out (especially if it is part of a design-out exercise). Table 3.1(a) describes the characteristics of the workload using this categorization. Such a categorization is of most use when evaluating the life plan, i.e. assessing how effective the preventive work is in controlling the level of corrective work. However, at this stage of the book we are concerned with the categorization of the workload as an aid to organizational design, so it is more sensible to do it as shown in Table 3.1(b) rather than as in 3.1(a), i.e. categorizing the work by its planning and scheduling characteristics (which is shown in more detail in Table 3.2). Table 3.3 shows an actual categorization for a few remotely located small power stations.

Table 3.1(a) Division by policy characteristics 3.1(b) Division by organizational characteristics

(a)	(b)
Corrective	**First line**
Emergency	Corrective emergency
Deferred	Corrective deferred (minor)
Workshop	Preventive routines (on-line)
Preventive	**Second line**
Routines (on-line)	Corrective deferred (major)
Services	Corrective workshop
Major (shut-down)	Preventive services
	Modification (minor)
Modification	**Third line**
Revenue (minor)	Corrective deferred (major)
Capital (major)	Preventive major (shut-down)
	Modification (major)

Table 3.2 Detailed categorization of maintenance workload by organizational characteristics

Main category	Sub-category	Category number	Comments
First line	Corrective-emergency	(1)	Occurs with random incidence and little warning and the job times also vary greatly. A typical emergency workload is shown in Figure 3.1. This is a workload generated by operating plant, the pattern following the production operating pattern (e.g. 5 days, 3 shifts per day etc). Requires urgent attention due to economic or safety imperatives. Planning limited to resource cover and some job instructions or decision guidelines. Can be off-line or on-line (in-situ corrective techniques). In some industries (e.g. power generation) failures can generate major work, these are usually infrequent but cause large work peaks.
	Corrective-deferred minor	(2)	Occurs in the same way as emergency corrective work but does not require urgent attention; it can be deferred until time and maintenance resources are available (it can be planned and scheduled). During plant operation some small jobs can be fitted into an emergency workload such as that of Figure 3.1 (smoothing).
	Preventive-routines	(3)	Short periodicity work, normally involving inspections and/or lubrication and/or minor replacements. Usually on-line and carried out by specialists or used to smooth an emergency workload such as that of Figure 3.1.
Second line	Corrective-deferred major	(4)	Same characteristics as (2) but of longer duration and requiring major planning and scheduling.
	Preventive-services	(5)	Involves minor off-line work carried out at short or medium length intervals. Scheduled with time tolerances for slotting and work smoothing purposes. Some work can be carried out on-line although most is carried out on-line during weekend or other shutdown windows.

Table 3.2 (cont'd)

Main category	Sub-category	Category number	Comments
	Corrective-reconditioning and fabrication	(6)	Similar to deferred work but is carried out away from the plant (second line maintenance) and usually by a separate tradeforce.
Third line	Preventive-major work (overhauls etc)	(7)	Involves overhauls of plant, plant sections of major units. Work is off-line and carried out at medium or long-term intervals. Such a workload varies in the long term as shown in Figure 3.1. The shutdown schedule for large multi-plant companies can be designed to smooth the company shutdown workload, see Figure 3.3.
	Modifications	(8)	Can be planned and scheduled some time ahead. The modification workload (often 'capital work') tends to rise to a peak at the end of the company financial year. This work can also be used to smooth the shutdown workload.

Table 3.3 Categorization of station workload by organizational characteristics

Work level	Category	Description
Typical first line work	A	Requires to be done within the shift in which it arises.
	B	Requires to be done within twenty-four hours of it arising.
	C	Minor corrective work that does not fall into categories A or B but does not require planning and is of relatively short duration.
	D	Minor routine preventive work, eg 500hr service that does not require a high degree of skill and can be carried out on a routine basis.
Typical second line work	E	Major corrective work that starts as a category A or B job.
	F	All corrective jobs that benefit from some form of planning and have a scheduling lead time > 24 hrs. Such jobs do not require a major influx of resources.

Table 3.3 (Cont'd)

	G	Modification work that has the same planning characteristics as category F.
	H	Preventive maintenance work that has the same planning characteristics as category F, eg this would include all services other than the major outages.
Typical third line work	I	Work that might involve considerable planning and scheduling effort in terms of job methods and major spare part resourcing. In addition, involves an influx of labour to resource peaks or has a specialist skill content.

Mapping the workload

One way of mapping a maintenance workload for a plant, plant area or trade group is shown in Figure 3.1, which shows the workload for the fitting group at a food processing plant (it was also shown in Book I). We shall refer to this figure, and Figure 3.2, throughout the following explanation of the general characteristics of this load.

First line workload: Mainly the emergency corrective work, jobs in this category have to be carried out immediately or within twenty-four hours – and are therefore impossible to schedule. At best, the average level of such work can be

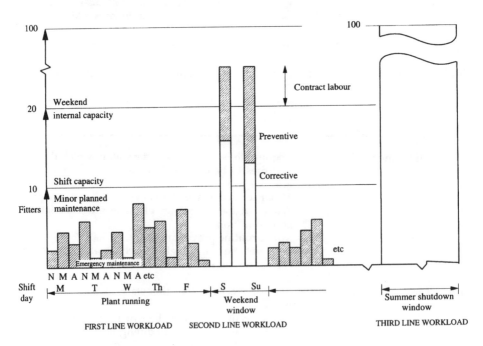

Figure 3.1 Workload pattern for fitters

forecast. Also classifiable as first line work are the simple deferred corrective jobs and the preventive routines (Categories 2 and 3 of Table 3.2), work which is often used to smooth the emergency workload (see Figure 3.1). It is this work, especially the emergency corrective, that determines the type and size of the first line resource. This is especially true of shift work where the plant operates over a full 24-hour day. It is important that this workload, and the operators' workload, be studied in depth before deciding on maintainer–operator flexibilities, inter-trade flexibilities and the adoption of self-empowered teams.

Second line workload: Consisting mainly of (a) the deferred corrective work that has a scheduling lead time of more than twenty-four hours, (b) the various services and (c) removed item work – Categories 4, 5 and 6 of Table 3.2 – these jobs are usually less than two days in duration and require relatively few tradesmen (often only one). The service work can be prioritized, planned and scheduled in the long term.

The deferred corrective work comes in on a continuous basis and needs to be prioritized, planned and scheduled. The second line plant resource groups need to be sized to handle the average input of this work plus the scheduled services. In the case of Figure 3.1 the average amount of second line work completed at weekends should equal the average weekly input of second line work, contract labour being used to peak lop. The situation can be visualized as in Figure 3.2. Clearly, the efficiency of the work planning system has a major effect on the size of the second line resource.

The removed-item work is in general carried out by a combination of contract reconditioning and internal workshop-based reconditioning or fabrication. The minor reconditioning work can sometimes be timed to smooth the second line workload.

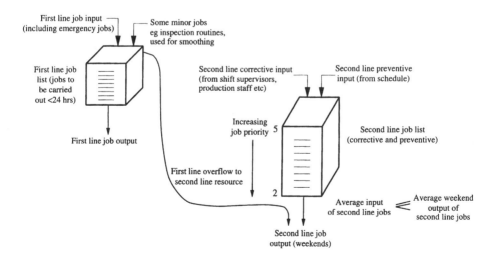

Figure 3.2 Visualization of the flow of maintenance work

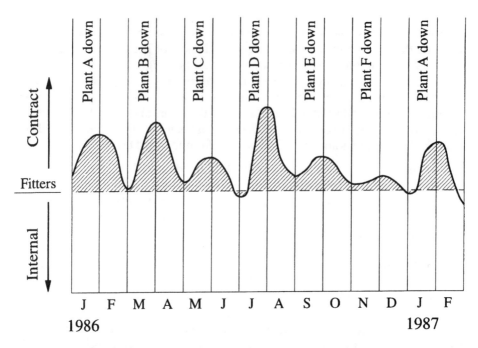

Figure 3.3   Shutdown workload, multi-plant company

Third line workload: Comprising the major shutdown and plant overhauls and any capital projects or modifications, its main characteristics are that it creates major peaks in resource requirement at medium or long-term intervals (and also involves many interrelated jobs that have to be completed in a specified time – typically in a few weeks). Plants that present a true third line workload of the type shown in Figure 3.1 are forced to bring in contract labour to supplement the internal resourcing of such peaks. Multi-plant companies (e.g. electricity utilities having several stations) can often arrange their schedule of major shutdowns so as to smooth their overall third line workload and hence minimize their requirement for contract labour (see Figure 3.3).

Forecasting the maintenance workload

Before modifying an existing maintenance organization, or designing a new one, the fullest possible information about the expected workload must be acquired. At a minimum, this should include estimates – for each major plant or area and for each trade, and for a representative and adequate period of time – of the following:

First line workload
(a) The emergency maintenance workload, i.e. the demand (and in particular, the maximum demand) in men per shift.

(b) The preventive routines and other major first line work, i.e. the average demand, in man-hours per day.

Second line workload

(c) The deferred corrective workload, i.e. the average demand in man-days per week. This should be further categorized by priority and by the plant status required for its execution (i.e. shutdown, alongside other work, etc.).

(d) The minor preventive (e.g. services) workload, i.e. the average demand in man-days per week. This should be further categorized by priority and by required plant status.

(e) The removed item and fabrication workload, i.e. the average demand in man-days per week (this would normally be amalgamated into a company-wide demand per week). There should also be an indication of whether the work should be contracted out and whether this should be to a centre of maintenance excellence.

Third line workload

(f) The expected major workload, i.e. the start-time, duration and size (in men per day) of each major overhaul (over a period of five years, say, for a power station). The workload diagram for each trade could be shown against the same time scale, enabling trade linkages to be indicated. In multi-plant companies evaluation of the long-term workload on a company-wide basis would facilitate workload smoothing. (See Figure 3.3, for example, which shows the third line workload for a chemical manufacturing company having several plants in the same complex. The various plant shutdowns were staggered throughout the year to smooth the workload. A resulting peak/trough ratio for the third line workload, of approximately 0.7 minimized the demand for contract resources).

Notes on workload forecasting

(i)   Forecasts of workloads (b), (d) and (f) above can be derived from the actual historical workloads (deduced from work order cards and/or management experience) and the future maintenance schedule.

(ii) Forecasts of workloads (a), (c) and (e) can also be based on the actual historical workloads but account must also be taken of the likely impact on the maintenance strategy – remembering that there is always a time lag before the corrective load responds to the preventive input. Forecasting for new plant is very much more difficult and must rely on management experience, manufacturers' information and experience of similar plant.

(iii) Unless some form of maintenance work measurement system is being used, e.g. comparative estimating[1], the estimated times in (a) to (e) will be based on what has gone before, i.e. will take no account of the organizational inefficiencies that may formerly have been present, and it will sometimes be necessary to make an allowance for this.

Case studies in categorizing and mapping the maintenance workload

1.   Agricultural chemicals

This shows how a large multinational company – manufacturing agricultural chemicals on a multi-plant site – categorized its workload and used this information in organizational design.

The company identified each job that made up the workload according to the criteria listed in the main column of Table 3.4. They then categorized each job in the right-hand columns into first, second or third line, with the following typical result:

First line	20%
Second line	60%
Third line	20%

They argued that the responsibilities for first line work, and for the necessary resources, were best carried locally, within each plant. In addition, the resource group for second line work should be shared between a number of plants and that for third line centralized or put out to contract. Because the total man-hours spent on each category were known, the maximum size for each group could be estimated. They further proposed that the nature of the fluctuations of the workloads in each category was such that the sizes of the first, second and third line groups should each be set to a minimum and that the work peaks should be allowed to cascade from first to second to third line and then to contract.

2.   Alumina refining

Illustrates how a workload profile can be mapped using information that is unavoidably limited and of low quality. The profile was established:

to obtain a feel for the performance and utilization of the tradegroups,
to assist an organizational redesign.

The plant concerned was a large, complex, continuously operating alumina refinery (discussed in detail in Chapter 4 of Book I). As a whole it never came off-line. Maintenance was undertaken at plant unit (e.g. bauxite mill) level because of the extensive redundancy which existed at this level. The off-line

Table 3.4 Criteria applied to work activities

Proposed criteria	First line		Second line	Third line
	Shifts	Days		
Small jobs<1 hour, small leaks, spanner jobs.				
Planning not required (straightforward job)				
Maybe a little organization required, <4 hours work.				
Tools, materials, technical information, joints.				
Need to be organized before being worked on.				
Requires day support to shift core.				
Preventive maintenance work-patrol/check list.				
Leaks, oil levels.				
Complex/multi-skill requiring planning.				
High frequency (per shift/day), greasing, oil checks, may be complicated.				
Low frequency (per week/month).				
Demands immediate response.				
Demands urgent response.				
Can wait for >1 week.				
Requires specialist skills.				
Requires local skills.				
Requires some plant skills.				
Requires specialist equipment/machinery.				
Requires doing in centre of excellence.				
Requires contractor.				
Workload shedding may be a problem				
Variety of work may be significantly affected.				
Have we split up what is an engineering transformation i.e. splitting whole tasks: multiple responsibility?				
Activities may be clustered.				
Any other criteria you feel may be necessary.				

work was therefore scheduled at fixed operational intervals in order to spread the workload throughout the year.

Because the plant was large many trade groups carried out the maintenance. The management felt, however, that specialized centralized groups needed to be set up to deal with the more sophisticated work. One such group was the mechanical drives group (MDG) who were responsible for the maintenance of the gearboxes, couplings and so forth of the kilns, mills etc. of the whole plant. This group worked only on the day shift and undertook the first line work, and also the second line plant work and reconditioning. The first line shift crew – a separate team – were only permitted to maintain this equipment in the event of an emergency. The author could see the advantages of the MDG specialization but felt it had gone too far. His view was that the MDG was underutilized.

As a first step to reviewing the situation the work profile of the MDG was estimated. It was constructed after (see Figure 3.4):

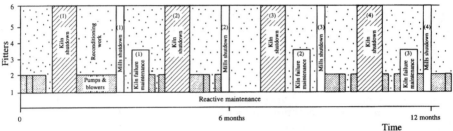

Figure 3.4 Estimated workload pattern for mechanical drive group

- reviewing the off-line preventive schedule and the histories of shutdowns;
- discussing the off-line failure histories with the tradeforce and supervisors and examining the work order history;
- asking the supervisors to estimate the average number of fitters on first line work.

The review showed that the workload was divided as follows:

First line maintenance	20%
Second line plant maintenance	40%
Reconditioning	40%

To a large extent the reconditioning work was being used to smooth the second line load, the overflow being contracted out.

The author was concerned about the following aspects of the arrangement:

- The first line work on mechanical drives should be carried out by the specialist shift team. If necessary their skills should be improved to enable them to perform this work satisfactorily.
- Much of the second line plant work could be carried out by the area plant mechanical teams, with specialist assistance as necessary from the MDG.
- The MDG should concentrate on carrying out quality workshop-based reconditioning. Even in this area, careful consideration should be given to contracting out work which would be better carried out at centres of excellence.

Reference

1. Kelly A., *Maintenance Planning and Control* (Appendix 4) Butterworths, London, 1984.

4
Maintenance resource structure

Introduction

The maintenance resource structure – various examples of which have been outlined in Figures 1.6, 1.7 and 2.2 – is concerned with the geographical location of personnel, tools, spares and information; their function, composition and size; and their logistics. In this chapter it will be shown how such a structure can be designed, modified and mapped – and how the key decisions that affect its shape and size can be identified.

Mapping the resource structure

This will be described via an example taken from a maintenance management audit of an alumina refinery (see Figure 4.1). Its various sub-processes were located as shown in Figure 4.2, a plant layout diagram which also identifies the locations of the trade groups. This ties up with Table 4.1 which shows the functions, compositions, size and shift rosters of the trade groups. The resource structure (see Figure 4.3) maps the trade groups by work function down the vertical axis (first line, second line, etc.) and by plant specialization or location along the horizontal axis. The operator groups are shown above the plant equipment line. For example, Group (a) is made up of thirteen fitters on days carrying out second line work in the grinding area. Other labour information –

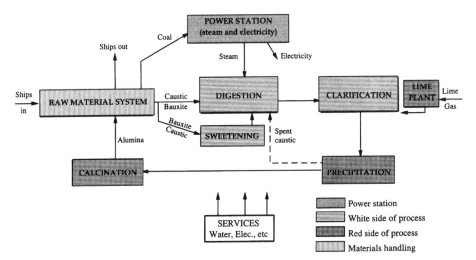

Figure 4.1   Alumina refinery, process flow

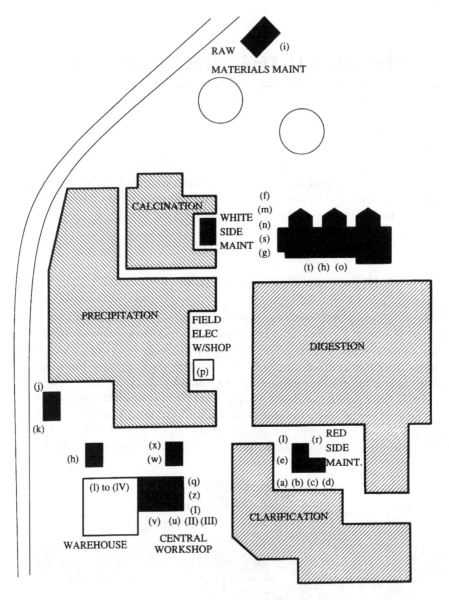

Letters indicate location of trade groups eg. (i)= Raw materials fitting group (see Figure 4.3)

Figure 4.2 Alumina refinery, plant layout

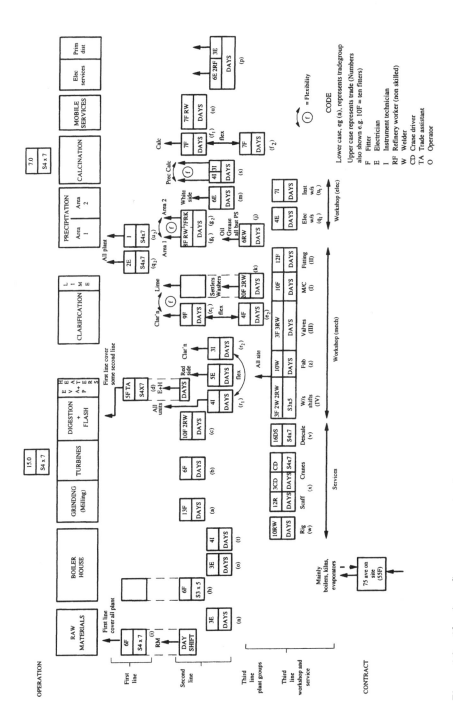

Figure 4.3   Alumina refinery, resource structure

Table 4.1 Extract from listing of trade group functions

TRADEGROUP NAME	LOCATION	COMPOSITION	SHIFT ROSTER	WORK FUNCTION
Raw materials mechanical	Wharf area workshop	24F	6F on a 4 x 7 roster	On Monday to Friday shift, material handling equipment, second line work. On all other shifts plant wide first line work, other than where there is local cover.
Raw materials electrical	Wharf area workshop	3E	Days	Material handling equipment, first and second line work.
Boiler house mechanical	Boiler house mechanical workshop	18F	6F on a 3 x 5 roster	First and second line cover for boiler house Monday to Friday

Table 4.2 Personnel inventory, alumina refinery

Salaried – Maintenance (1992)
Engineering managers   12
Engineers (including allowance
for plant engineering support)   17
Supervisors   Direct   43
                Planning   9
                Training   4
                Total   56
Clerical   4
Total staff   89

Waged inventory
Trades
                Fitters   194
                Welders   16
                Electricians   35
                Inst. tech.   29
Total   274

Non-trades (RW)
                Trades assistants   22
                Lubrication   6
                Crane drivers   7
                Scaffolding   12
                De-scale and others   74
                Total   121

Total waged   395

Simple ratios

Trades/non-trades+   8.0
Waged/salaried   4.4
Waged/direct supv   9.2
Waged/planning supv   4.4
Waged/total supv   7.1

e.g. on inter-trade flexibility, inter-plant flexibility, human factors (ownership, goodwill, morale) – is best recorded in the description of a figure such as 4.3 or in a table such as 4.1. A complete labour inventory is shown in Table 4.2, which also uses simple indices to identify some important characteristics, such as trades/non-trades ratios.

The locations of the spares and tool stores are best shown on the plant layout diagram.

Defining the resource structure's decision problems

The design of a resource structure should aim to achieve the best balance between the utilization of the tradeforce and the quality of service it provides. In practice, this aim will need to be defined for each situation and will most likely be to achieve the best resource utilization for a desired response and work quality. Although such an objective is difficult to quantify it can be used to judge the relative merits of possible modifications to the structure, or of alternative structure designs (see Figure 4.4).

What the basic problem of the resource structure boils down to is deciding on the best way to match the maintenance resources to the maintenance workload. As can be seen from Figure 4.3, this is a complex task involving many decisions. There are a number of key questions, however, that must be addressed at the start, e.g. 'Where should the resources be located?' 'Should we use maintainer/operator flexibility?' 'Should the internal resources be centralized?' etc. Each decision will be influenced by many factors – the main ones being the *workload* characteristics and *resource* characteristics. The former have been discussed in Chapter 3 but the latter will need to be reviewed before going on to a detailed examination of the nature of the key decisions.

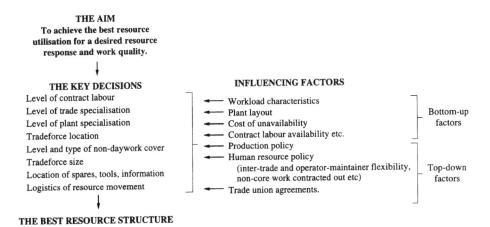

Figure 4.4   A model of the resource structure decision problem

Resource characteristics

Manpower. This may be classified according to the technical area in which it is employed (mechanical, electrical, instrumentation, building, etc.), further divided according to craft (fitter, welder, electrician, etc.) and, if necessary, sub-divided into specialization (boiler, fitter, turbine fitter, etc.).

The quality of labour available will depend mainly on the environment within which the company operates, on the technical and craft training system, on the availability of retraining and specialist training, on the availability of contract labour and on the influence and attitudes of trade unions.

The factors that influence the morale and motivation of the maintenance workforce are similar to those influencing other shop floor workers. However, the maintenance tradesman is one of the few shop floor workers who still has considerable autonomy over his day-to-day actions and decisions. In addition, maintenance work has many of the attributes that promote worker satisfaction – craftsman status, pride in the quality of the work, varied and interesting job content. An important human factor at shop floor level is the sense of *equipment ownership*, which can apply to both operators and maintainers. Of increasing importance is *team working* – especially between operators and maintainers. Perhaps the most important human factor is goodwill towards the company – a characteristic which is not easy to promote via activities operating at shop floor level only. The nature of these various factors differs enormously from one country, or one company, to another and the local situation regarding them must be fully understood before any organizational re-design is attempted.

Spare parts. The objective of spares organization is to achieve the optimal balance between the cost of ordering and holding (depreciation, interest charges, rental, etc.) and the cost of stockout (loss of sales due to unavailability, temporary hire charges, etc.). The main difficulty in this simply stated task arises from the variety and complexity of the many thousands of different items (of widely varying cost, lead times and usage rates) required to sustain a typical operation. Each spare part requires its own inventory policy and, in a sense, presents an individual problem of control.

To facilitate the setting of control policies spare parts can be classified as either *fast moving* (demand per year greater than three) or *slow-moving* (three or less). The slow movers can be further classified as *adequate warning* or *inadequate warning* items. The slow-moving inadequate warning category often accounts for a large proportion of the stockholding costs (see Chapter 11). To further facilitate their management, spares should also be classified, of course, according to their function (abrasives, bearings, etc.).

Companies in the developing world (or remote from suppliers) often experience long and variable lead times that cause considerable difficulty for the setting and controlling of spares inventories – which can then cause problems in the execution of maintenance strategy. Options for local reconditioning of parts should then be a major consideration.

With regard to spares, the main resource structure decision concerns the function, location and number of stores. In the Figure 4.2 example there was a single centralized stores, the main computer-controlled warehouse. In addition, each main workshop (highlighted in the figure) had a minor-parts store for consumables and other non-computer-controlled items. The danger with such sub-stores is that the local supervision can build them up in competition with the main stores ('magpie-ing'). If this gets out of hand it can destroy the main system. An associated secondary problem is the logistics, i.e. the management of the movement of spares – including that of component parts and 'rotables' (see Figure 4.5).

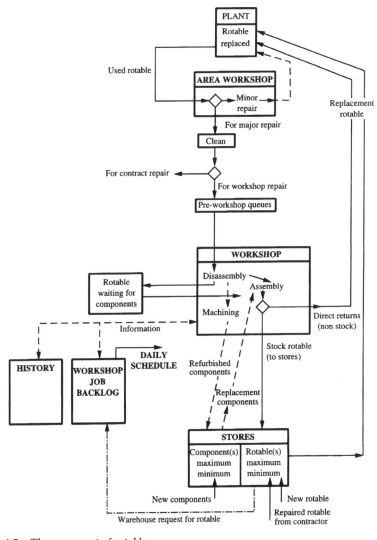

Figure 4.5   The movement of rotables

Tools. Although the objective of the organization of tools is similar to that of spares the control problem is different because tools are not, in the same sense, consumable. Tools, like spares, can be categorized, the simplest division being as follows:

small	(sometimes supplied to individual tradesmen);
large	(held, for issue, in a tool stores);
lifting gear and scaffolding	(held separately and usually subject to periodic testing);
electronic	(held separately and often in a controlled environment).

The main task with returnable tools is the development of a system for monitoring their loan and maintaining them (or replacing them if necessary) when returned. A simple tool control system is shown in Figure 4.6.

Information. Included under this heading are all documents, catalogues, manuals or drawings that might facilitate maintenance work. They fall into several categories:

- *Training.* Used primarily for initial training of the tradeforce on new equipment (e.g. logical fault-finding manuals).
- *Reference.* *Might* be consulted before carrying out a job (e.g. manuals, item histories, spares catalogues).
- *Instruction.* *Must* be consulted before carrying out a job (e.g. work orders, compulsory job instructions, safety instructions).

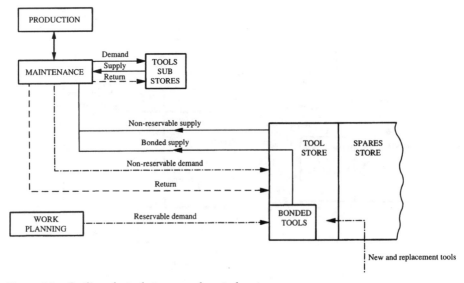

Figure 4.6   Outline of a tool storage and control system

- *Control.*   Used to record and control the way in which work is carried out.

Each of the above categories has different characteristics. Some of the questions that need to be asked are:

How is the information to be held?   (on paper, on computer or on some combination of these)?

Where is it to be held?   (centrally, locally, or in some combination of these)?

Who is to use such information?   (the tradeforce or the manage-ment)?

This kind of inquiry is particularly important when building up a user-requirement statement when procuring a computerized maintenance information system. The questions cannot be answered effectively without reference to models of the kind shown in Figure 4.3.

The key decision-making areas of resource structuring (see Figure 4.4)

Contract labour. Deciding on the extent to which outside labour is used. Traditionally, contract labour has been used for the following reasons:

(i)   *Resourcing peaks in labour demand (mainly for third line work).*   Improves the planning function's ability to match resources to a fluctuating workload which, in turn, improves overall labour utilization. The disadvantages include: slower response than that of internal labour (the larger the peak the greater the lead time for resourcing it); lack of plant knowledge; lack of identification with the company; and increased hourly labour cost. These can be only partially compensated by detailed job contracts and internal supervision of work quality.

(ii) *Specialized work.* The advantage of employing external labour increases as maintenance work becomes more sophisticated and specialized – e.g. gas-turbine maintenance – and, as is often the case with such work, more peaky in its workload pattern.

(iii) *Reconditioning units or assemblies.*   This has considerable advantages where sophisticated equipment is needed for reconditioning or quality checking and where such equipment would not be fully utilized internally. The disadvantages can be long reconditioning lead time and high cost.

In the resource structure of Figure 4.3 contract labour is used for reconditioning the more complex or sophisticated rotables, e.g. large gearboxes, where special equipment or expertize is required. In addition, there is a

Table 4.3 Criteria used to identify a core maintenance service

(i)   Critical to one or more of the business units on the site.
(ii) Provides a rapid skilled response to cover emergency maintenance.
(iii) Provides a specialized resource or skill which is not readily available from outside contractors.
(iv) Involves many short duration jobs scattered throughout a wide area.
(v)   Local plant or site knowledge is needed.
(vi) Close interaction with operations or with the users of the service is essential.

pool of contract labour continually on site which varies in size depending on the demand. In this case the contractors smooth out the second line work peaks (there is no real third line work) and are allocated the 'dirty' jobs. Generally speaking, the pool is used as a sink for jobs that are not wanted by the internal labour force. Most medium or large-sized companies have customarily used contract labour to supplement their own (often large) internal labour forces.

Recently, criteria have been developed for identifying the maintenance work, the core work, that can be considered essential to the companies main function. Table 4.3, for example, lists the criteria laid down by Riddell during his reorganization of the maintenance department of a large chemical company. He argued that a contractor can better supply the resources for all non-core work, relieving the company of the associated man-management and industrial relations problems so that it can concentrate on its core business. Such a policy has been used for many years in the maintenance of large building complexes.

For the estimated workload shown in Figure 4.7, for example, the proposed organization was as outlined in Figure 4.8. Although there was a need for a resident maintenance group for first line work it was considered that such work was non-core and it was made the contractual responsibility of an outside company.

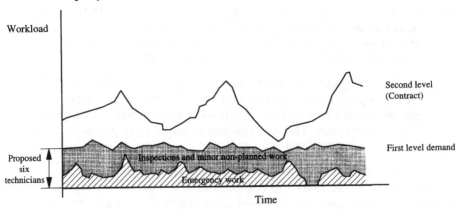

Figure 4.7   Estimated workload for a new building complex

Table 4.4 Proposed guideline for establishing maintenance workload

Main category	Sub category	Cat no.	Job description	Trade(s)		Core work	Non-core
				Main	Sub		
First line	Emergency	(1)					
Second line	Deferred corrective etc	(4)					
Third line	Major overhaul	(7)					
		etc					

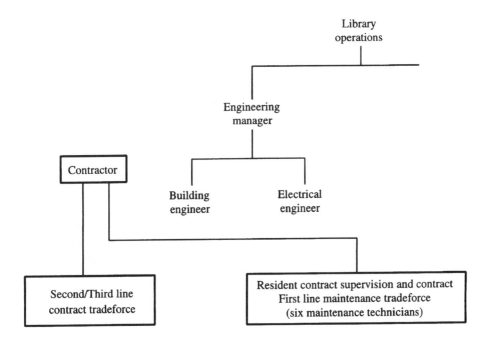

Figure 4.8   Proposed maintenance organization for a new building complex

Table 4.5 Some of the principal disadvantages of using contract labour

- Difficult to promote plant ownership.
- Poor plant knowledge requires detailed instructions and information.
- More difficult to obtain plant maintenance information for history recording.
- Contractors have different objectives from the company and are not overly concerned about the longer term condition of plant.
- Conflict can arise between company and contract tradeforce.
- Company safety procedures may not be adhered to. Poor quality of maintenance work can often go unnoticed and might cause plant safety problems.
- Need to use several contract firms to avoid being 'held to ransom'.

As explained, the above contracting-out was not done for specifically financial reasons, but to enable the management to concentrate on its main function. The contracting company, however, still had to make many of the resource structure decisions for the non-core work – regarding such matters as composition and location of the tradegroups, flexibility, shift working, etc. The adoption of this policy means that the decision to contract out becomes the most important one in the design or modification of the resource structure. It follows that the categorization of the work load shown in Tables 3.2 and 3.4 should be extended to the identification of core and non-core work. This might be accomplished using a format of the kind shown in Table 4.4.

The more important *advantages* in using contract labour have been discussed. The principal *disadvantages* are listed in Table 4.5.

Tradeforce composition. Identifying the various trades, operator grades and unskilled grades to be used; this includes defining work roles, skills and the degree of flexibility between trades, between tradesmen and operators and between tradesmen and unskilled workers. In general, the greater the division of work (boiler fitter, turbine fitter, electronic technician, etc.) the greater the skill of the individual trades.

In the UK and many other countries it has been customary to divide the maintenance workforce into various trades and sub-trades and also into several categories of unskilled worker. Each individual trade has been union-protected and its skills highly demarcated. This lack of flexibility causes inefficiency in the planning of multi-trade jobs.

In terms of maintenance resource design the question of trade specialization and flexibility can be explained via reference to Figure 4.9. Where the work requires skill and where the workload can be made relatively smooth there are considerable advantages in trade (e.g. high pressure welding) specialization. This is the situation that often exists in the second line reconditioning workshop. On the other hand, the second line plant maintenance groups more often than not carry out jobs that require a range of skills, although one skill is usually predominant. In these situations inter-trade flexibility is of paramount

THIRD LINE MAINTENANCE
Workshop reconditioning
and fabrication,

SECOND LINE MAINTENANCE
Area in-situ plant
maintenance

Servicing

Overhauls

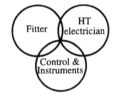

FIRST LINE MAINTENANCE
Emergency repair.
Inspections and minor routines

Figure 4.9   Work characteristics and trade skills

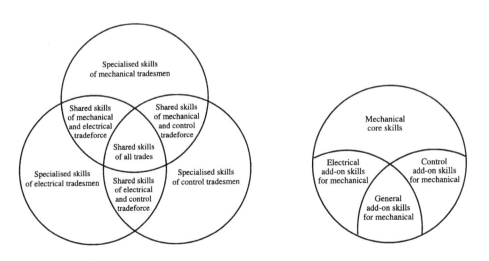

Figure 4.10   (a) Inter-trade flexibility (b) Skills needed by mechanical tradesmen to facilitate flexible operations

importance, i.e. retention of the basic core skill – fitting, say – with 'add-on' skills in other areas to facilitate inter-trade flexibility (see Figure 4.10). These are, however, only very general observations; the author has encountered situations in oil refineries where the engineering manager was adamant that he still needed specialization for second line plant groups. In the electrical and electronic area, for example, he felt that the following technicians were needed:

Specialist	Generalist	Specialist
HT electrical	Electrical/instrument	Electronic/instrument

In some industries there are considerable advantages in amalgamating the roles of operator and first line maintainer, creating an operator who undertakes a range of first line work (electrical, mechanical or electronic) for a small area of plant or process stream. When applied to shift workers this can lead to considerable labour productivity gains. One common way of achieving this is to recruit tradesmen as operators.

In the example of Figure 4.3 considerable training had improved the range of skills of the individual tradesmen. However, because of industrial relations problems the inter-trade flexibility was restricted to the consolidation of the fitting skills into a single mechanical tradesman ('F' for fitter in the figure). In addition there was limited flexibility across the interfaces operator/mechanical, electrician/instrument technician and fitter/electrician.

Labour flexibility, particularly with regard to simpler work, can be improved in the longer term in two principal ways. First, through productivity agreements and other management–workforce bargains, thus reducing demarcations (see Figure 4.9), e.g. the mechanical tradesman could engage in general fitting, welding, cutting, metal forming, etc. Secondly, through formal training programmes, both internal and external, to extend the tradesman's skills. The benefits of improved flexibility include easier work planning and higher utilization. This, however, has to be weighed against the costs of the productivity deal, of training and of the installation of the scheme.

In the short term the resource structure has to be designed, or redesigned, within the constraints of the existing flexibility agreements between management and trade unions.

Plant specialization. Deciding the extent to which the tradeforce is dedicated to the maintenance of a single plant, area or unit type. In a large company the extremes can range from a *plant-flexible* tradesman who is expected to work on all plant to a *plant-specialized* one who only works in a single area or on a particular equipment type, e.g. only on compressors.

The advantage gained through plant specialization is improved work quality through greater plant knowledge and sense of ownership. This is especially the case where tradesmen are delegated the authority to control their own work. To a certain extent there is also a better response to job requests. Plant specialization also lends itself to the setting up of plant-oriented teams (POTs), i.e. of maintainers and operators jointly responsible for an area or section of plant (on days or shifts). This in turn may lead to the creation of the operator-first-line-maintainer described in the previous section.

The main disadvantage with plant specialization occurs where there is a peaky workload or where a considerable degree of work planning and control is required to smooth the workload. In such situations it is difficult to achieve high labour utilization. This is most evident when there are numerous plant-specialized groups each made up of highly demarcated single trades. This is especially true if a proportion of the groups are shift working. By and large, this is the situation of the Figure 4.3 example. The structure is traditional and has evolved over many years into numerous single trade groups. There are three first line groups and 22 second line plant-located groups. In addition there are eleven workshop groups. It is inevitable here that the tradeforce utilization will be low and work planning particularly difficult. Thus, if the advantages of plant specialization are required (and they are becoming increasingly important because of the trend towards plant ownership and team-working, and because of the increased focus on assuring product quality) then it must be combined with flexible working practices and, where possible, self-empowerment.

The benefit of inter-plant flexibility is the increased ability to balance a peaky workload by exploiting labour mobility. This makes for higher labour utilization but is usually accompanied by loss of plant knowledge and ownership. The former disadvantage is often mitigated by using extensive job instructions and the latter by making professional and technician engineers responsible for particular plant areas.

Tradeforce location. Deciding, for example whether workshops should be centralized or dispersed (a decision which tends to be connected with plant specialization decisions). Plant-flexible trade groups are normally located centrally. The plant-specialized area groups can be located centrally (or in a central workshop area) or close to their designated area (which is usually the case). The location of a trade group specialized in the maintenance of an equipment type will clearly depend on the distribution of such equipment.

In addition to the above possibilities groups can also be designated as 'roving'. This is especially suitable where a particular unit type is in use at widely scattered locations (e.g. compressors in a large oil field).

The main benefits of decentralizing the plant-specialized groups are increased speed of response and a stronger sense of team-working with the operators in the locations concerned. This is especially the case if Production and Maintenance both report to the same authority. If plant-oriented teams are to be developed the plant-specialized maintainers must be located near to their plant and to the plant's operators.

The main disadvantage of decentralized plant-specialized groups is the difficulty of achieving flexible labour movement between trade groups (this is even more difficult if plant-oriented teams have been adopted). In theory, inter-plant flexibility can be used to move labour from groups with a low workload to those with a high one. In practice, this is particularly difficult because of human factor reasons. People do not like moving out of their groups and do not like accommodating strangers within their group.

Non-daywork maintenance cover. Deciding the way in which maintenance activity outside the normal daywork is resourced (i.e. outside, say, 08.00 to 17.00 h Monday to Friday). Possibilities include shift working, staggered day-shifts, overtime, call-out, or combinations of these.

The need for non-daywork maintenance arises mainly for the following reasons:

- There may be a demand for emergency maintenance cover for plant operating outside normal daytime hours.
- Planned off-line maintenance may be scheduled to be undertaken outside normal production times or during a major overhaul.

In the case of non-daywork emergency cover the advantages of shift working are rapid response and the fostering of team spirit between production and maintenance shift workers, if their shifts coincide. The disadvantages are the usual ones associated with decentralization, but magnified by the increase in the number of small shift groups – a situation in which supervision and planning is inherently difficult. Utilization can often be improved by centralizing shift cover and using a call-out system. The greatest gains in utilization in this area, however, have come about by increasing flexibility across the interface between the first line maintainer and the operator (in some cases shift operators have been trained to carry out essential first line maintenance). With the resource structure of Figure 4.3 all first line maintenance is carried out by shift groups, because the plant is in continuous operation; in addition there is a call-out system for key personnel. In this case it would appear that there could be benefits from combining the groups into a centralized first line shift group located in a well-equipped workshop.

For non-daywork scheduled activities the advantage of shift working lies in the quicker completion of major overhauls, which is particularly important where downtime is costly (e.g. in power generation). The disadvantage is in the difficulty of co-ordinating the work between several different groups of workers. A further problem is caused by the peaky nature of such work. Shift working flexibility is therefore needed if high labour utilization is to be achieved – i.e. when the shift workload drops, the centre of gravity of the shift roster should move towards day working.

The more usual way of resourcing a peaky work load such as the above is through an overtime arrangement. In most cases this has proved to be the most flexible and economic solution. In many situations, however, the use of overtime has been abused. It has become a means of supplementing low basic wages and the tradeforce has come to expect overtime as of right. Jobs in normal daywork are deliberately slowed to extend into the overtime period; those, such as major shutdown work, that are to be carried out on overtime are deliberately prolonged. This human factor problem has been overcome in many companies by the introduction of *annualized-hours agreements*. An agreed

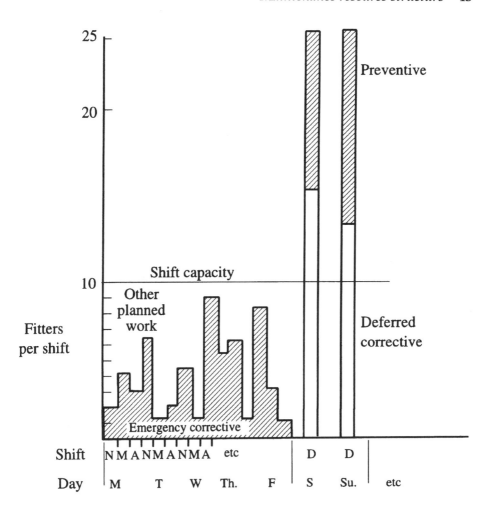

Figure 4.11   Food processing plant, Monday to Friday workload (fitters)

annual level of overtime payment is awarded in 52 weekly instalments, regardless of the actual hours worked. The burden of the implementation of the scheme rests firmly with the tradeforce, the agreement often being bought-in as part of a self-empowerment programme. The tradeforce in turn agrees to get the work done as quickly as possible. Actual overtime is reduced and plant availability improved.

The main point that needs to be stressed is that resourcing non-daywork cover has to be considered in relation to workload pattern, work roster agreements, cost of unavailability, required speed of response, etc. The experience of the food processing plant of Chapter 3 illustrates this very well. The company felt that the most economic way of covering the

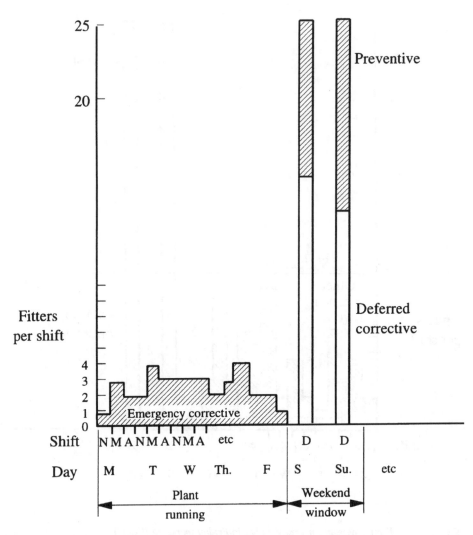

Figure 4.12   Biscuit factory, Monday to Friday workload (fitters)

Monday to Saturday workload (see Figure 4.11) was to use shift working from Monday to Friday and then to use the shift workers on overtime to carry out the second line weekend work. An alternative would have been some form of shift roster to ensure that at least half of the tradeforce were in on Saturdays and Sundays. More recently the author has audited a biscuit factory with a similar workload (see Figure 4.12). In this case the company used a shift roster to maximize the size of the tradeforce at weekends.

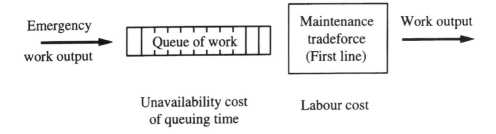

Figure 4.13   Simple queuing model for emergency maintenance

Sizing the tradeforce. Deciding the number of maintenance tradesmen allocated to each workshop or plant area. This number is a function of the workload, of the motivation of the men and of the efficiency of work planning (see Chapter 8).

The most complex and difficult sizing problem concerns the trade groups responsible for the high priority – mostly emergency – jobs. This is particularly the case where the tradegroups are working on shift rosters (see, for example, the first line shift groups of Figure 4.3). The duration of such jobs may be quite unpredictable and their incidence quite random but they will need to be undertaken within the shift in which they occur. This results in a workload with short duration intense peaks as in Figure 4.11.

A simple queuing theory model of the above situation is illustrated in Figure 4.13.* The objective is to identify the trade group size which minimizes the total cost of unavailability and labour. If the size is set at a level well below that needed to meet peak demand then the cost of labour will be low but the cost of unavailability high. Conversely, if it is set at a level above that needed to meet peak demand then waiting cost will be zero but labour cost will be high. Clearly, the main factor is the cost of unavailability. In many modern industrial plants the cost of unavailability is of a much higher order than the cost of labour so any peaks in repair demand caused by emergencies must be covered. This will inevitably lead to low labour utilization unless compensatory measures are taken.

The principal alternatives are:

(i)   workload smoothing by the inclusion of work of a lower order priority (see Figure 4.11);

(ii)   using inter-plant labour flexibility to resource the peaks (there will, of course, be an inevitable penalty in the shape of the unavailability costs caused by travel time);

(iii)   allowing the peaks to cascade to the second line groups (supplemented by call-out in the case of shift work);

(iv)   centralizing emergency cover, which will improve utilization –

---

* Queuing theory is discussed in Chapter 6, *Management of Industrial Maintenance*, A. Kelly and M. J. Harris, Newnes-Butterworths, London, 1978.

because the workload will be smoother – but has the disadvantage of slower response.

These alternatives can be used singly or in combination.

If the workload can be planned and scheduled (e.g. as it was for the second and third line trade groups of Figure 4.3) then for a given level of tradeforce performance it is not difficult to decide on the best size of trade group. If such a workload changes in the longer term, inter-plant flexibility can be exploited to smooth the peaks. Alternatively, the peaks can be allowed to cascade to a third line tradegroup (as existed in the Figure 4.3 example) or to contract labour.

Location of spares, tools and information. Identifying an optimal or, at least, an effective positioning of these resources is a task which is secondary, but closely related, to that of locating the tradeforce.

Decentralization of the tradeforce creates the need for decentralized sub-stores for tools and parts, in order to facilitate rapid response. This has to be considered against the accompanying disadvantages of increasing the costs of holding and of stores administration (especially if shift work is involved). If attention is not given to the administration of the sub-stores, slack control will lead to duplication of parts holding on the one hand and to more frequent incidence of stockouts on the other.

The need for decentralized information holding will also arise. Although information can indeed be copied and held in different areas it is essential to have a master information base (the transfer of information from which has been made much easier by the advent of computers).

Logistics. Deciding how the resources are to be moved around the site, e.g. planning the movement of repairable parts between plant, workshop, external contractor and stores (see Figure 4.5). These decisions are secondary, but closely related, to those of resource location.

A systematic procedure for determining a resource structure

It will be appreciated from the preceding discussion that the modification of an existing resource structure or the design of one for a new plant are complex tasks. With new plant there will be considerable uncertainty regarding the influencing factors, in particular regarding the size and pattern of the workload, which at best can only be estimated (with some allowance made for its increase as the plant ages). Where a resource structure has been in existence for a number of years there should be a recorded and detailed history of both workload and resource, and if there is a problem it is usually some mis-match between them. Poor management or artificial constraints on decision making may have led to a considerable excess of resource above the level needed to deal with the true workload. Clearly, modifying an existing resource structure and formulating one for a new plant are tasks which call for very different approaches, proposals for which are now outlined.

For a new plant

1. Understand the resource characteristics, giving particular attention to those that are particular to the situation under study.
2. Draw a plant layout and estimate the composition, size and pattern of the workload arising for each plant and trade. This should be categorized into first, second and third line, core and non-core (see Table 4.4).
3. Determine the level of response which production will find acceptable for emergency maintenance in each plant, area and unit.
4. Formulate the maintenance tradeforce structure, as follows:
    (a) *For each trade, outline an initial first line structure which will meet the response requirement for the anticipated emergency work.* This will involve the determination of numbers, locations and shift rosters. Consider possible associations of trades, and of maintenance and production labour, to form plant-based or product-based teams. Estimate the probable level of scheduled minor work, such as lubrication routines, that can be carried out by the first line tradeforce during normal hours (and which can be used for work smoothing). *Keep the first line trade group as small as possible commensurate with the workload,* using 'cascade to second line' and 'call-out' arrangements as necessary.
    (b) *For each trade, formulate an initial second line structure which will meet the anticipated ongoing scheduled workload but not major shutdowns or overhauls.* Such a structure might consist of an independent centralized pool or, as in the example of Figure 4.3, several small decentralized trade groups. In the food processing plant of Chapter 1 it involved setting up a weekend trade group on overtime, drawn from the Monday to Friday first line groups. *Again, the tradeforce size should be kept to the minimum commensurate with the estimated workload,* any peaks being resourced by cascade to third line or contract, whichever is available.
    (c) *For each trade, formulate an initial second line structure which will meet the expected reconditioning and fabrication workload,* taking into consideration possible contributions from the other work groups or opportunities for contract reconditioning.

(d) *For each trade, outline an initial third line structure which will meet the planned major workload and, in some cases, the anticipated unplanned major outages.* The most usual situation is that there is a need to resource major work peaks arising during complete plant shutdowns which only occur at long intervals. This will necessitate restructuring the internal labour force into a project type structure, allocating tradesmen according to their knowledge of particular plant. Where necessary, the level of contract labour required to supplement the internal labour force should be estimated and a contract procedure established which will ensure the supply of such labour when needed (e.g. preferred contractors might be identified).

In large multi-plant companies the need might be to resource a continuing, but fluctuating, major workload (see Figure 3.3), in which case a permanent major-work trade group could be established. The existence of such a group might have a considerable effect on the rest of the structure. The group should be sized to meet the trough of the workload, the peaks being resourced via contract or overtime.

(e) Using the information obtained during steps (a) to (d) formulate a proposed *complete* tradeforce structure. This formulation should rationalize the initial structures by exploiting, wherever possible, opportunities for flexibility – inter-trade, inter-plant, shift-related or overtime-related. The structure should be arranged so that:

(i) work cascades, in the ongoing structure, from first line to second line to contract;

(ii) labour cascades from the ongoing work to weekend shutdowns to major plant shutdowns.

Because of the uncertainty in the size and pattern of the workload for new plant it is prudent to restrict the size of the permanently employed tradeforce until experience has been gained.

Note. The way non-core work is handled will depend on the criteria used for its definition. At the one extreme it might mean work not directly concerned with production, i.e. maintenance of building services (painting, plumbing and so on). This might involve a small resident contract team with appropriate supervision. The company specifies the work (via its maintenance department) but the contractor is responsible for the labour and resource structure decisions. At the other extreme non-core work might be defined as all maintenance other than first line. In such a situation the company would have to set up a contract administration to specify the work and monitor its quality. Once again, however, the contractor

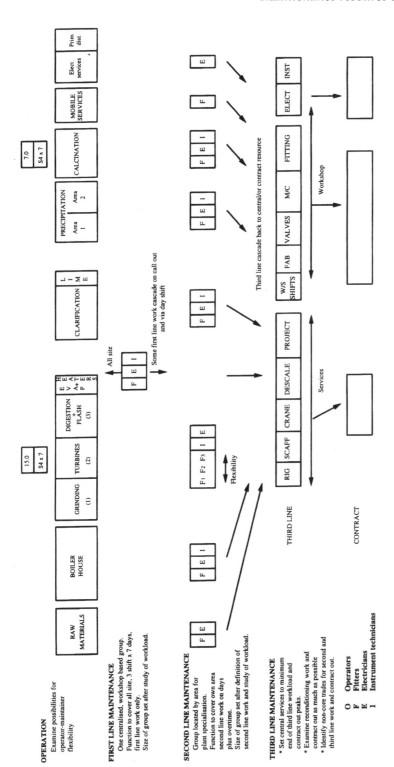

Figure 4.14 An approach to improving the resource structure

is responsible for the labour and the resource structure decisions (see for example Figure 4.8).

5. Formulate the spare parts and tools structure, taking account of the tradeforce structure established in Step 4.
6. Formulate the resource logistics for the combined structure established in Steps 4 and 5.

For an existing resource structure

The previous procedure has to be modified. Steps 1 to 3 are changed, because the existing workload and resource structure have to be mapped. The result of such an exercise – carried out as part of a maintenance management audit of an alumina refinery – was shown in Figure 4.3. Other information from the audit is shown in Figures 4.1 and 4.2, and in Tables 4.1 and 4.2.

The mapping would include not only the information outlined in Steps 1 to 3, but much more. For example, higher management may have decided that non-core work should be identified and contracted out, and also that shop floor 'ownership' should be fostered by amalgamating the maintainer and operator roles. The mapping may also reveal that the company is over-resourced – perhaps because its operations have been cut back but its maintenance resources have not.

The information obtained as a result of the mapping can be used in conjunction with the procedure of Step 4 to modify the existing structure in order to improve its effectiveness. The structure shown in Figure 4.14 was a proposal which resulted from conducting such an analysis on the structure shown in Figure 4.3. It was accompanied by the following observations.

(i)   The first line work should be clearly defined and an estimate made of its pattern and size throughout the refinery. Using this information, it might be possible to determine the size of a centralized and workshop-located shift crew (mechanical, electrical and instrumentation) to replace the several existing sources of shift cover. Such a first line crew should also carry out first line work during the day-shift allowing, as far as possible, the second line area crews to carry out the scheduled second line work. With suitable training, more of the first line work might well be carried out by the operators.

(ii) The second line work should be clearly defined. It would appear, for example, that a distinction can be made between the ongoing schedulable work and the 'turnaround' work (unit overhauls). The point being made is that it is necessary to distinguish the work (mechanical, electrical or instrumentation) that is best carried out by an area resource (for reasons of specialized knowledge, response, team-work) from work such as overhaul that is best carried out by a centralized resource. To a certain extent this is already done by the project group of Figure 4.3. Using this reasoning, the area second line groups should be sized and the excess manpower (if any) moved back to form a centralized third line plant resource to supplement the existing project group. This third line group should be set at the minimum size and peaks in its workload should be cascaded to contract.

(iii) Coupled with better production scheduling, adopting the above approach would mean that, initially, improvements in productivity would be gained by reducing the use of contract labour and not by reducing the company labour force.

(iv) The central workshop workload was not analysed either as regards its size or as regards the basis adopted for deciding whether reconditioning and fabrication should be undertaken internally or externally. It is recommended that such an analysis should be done and as much work as possible should be contracted out. The impression, a purely subjective one, has been gained that much of the work carried out within the workshops could be handled by outside contract. The main point here is that support activities of this kind should not be allowed to gain a momentum of their own (the company's business is alumina refining not engineering).

Summary

The design or modification of a resource structure is a complex problem involving a number of interrelated decisions (see Figure 4.3) some of which affect the shape of the structure (e.g. location, plant specialization) and some its size (e.g. trade specialization). What we are seeking is the best shape and size to match the distribution and pattern of the maintenance workload. Some of the key points to be borne in mind when carrying out this task are the following:

i) Although there is no one best type of solution the cascade structure, as in Figure 4.14, tends to suit the general characteristics of the maintenance workload.

(ii) If dynamic matching of the tradeforce to the workload is to be achieved, *flexibility* is the most desirable characteristic to be fostered in the tradeforce, i.e.

inter-trade flexibility,
maintainer-operator flexibility,
inter-plant flexibility,
flexibility of location,
shiftworking flexibility,
flexibility to use contract and/or temporary labour.

(iii) The design of the resource structure must always take into consideration the effect of the nature of that structure on the administrative structure and on work planning. It is only one part of the organization (see Figure 1.5).

5
Maintenance administrative structure

Introduction

The administrative structure, one of the principal elements of the maintenance organization, is a complex of managerial roles for deciding when and how an industrial plant should be maintained (see Chapter 2, where it has already been defined and illustrated). It differs from a resource structure in that the latter is concerned with the composition and location of the resources, whereas the administrative structure is concerned with allocating the responsibility for carrying out the work. Its principal functions are:

- the initial formulation and ongoing modification of the maintenance objectives, strategy, organization and control (including resource budgeting);
- the management of the maintenance resources (a necessary part of which is the transmission of the objectives, policy decisions and other information from senior management to tradesmen).

Modelling administrative structures

One way of modelling an administrative structure is to use an organization chart (or 'organogram'!) in which position titles are located so as to show their various responsibilities and lines of communication. Each title can be supplemented by a full position description, and an organizational manual can clarify the relationships between the various roles.

An example of such a structure, used for administrating the alumina refinery maintenance resources (see Figure 4.3), is shown in Figures 5.1 and 5.2.

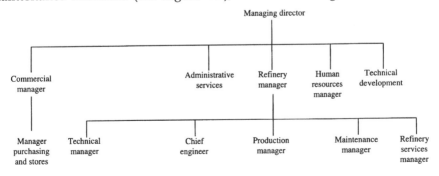

Figure 5.1   Senior administration, alumina refinery

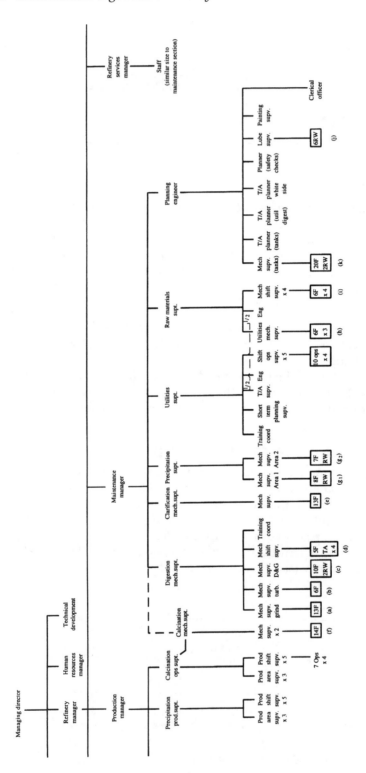

Figure 5.2   Extract from maintenance administrative structure, alumina refinery

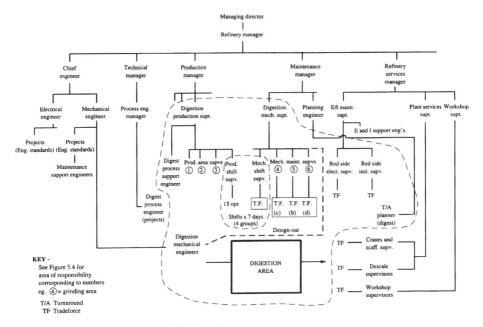

Figure 5.3   Digestion process administration

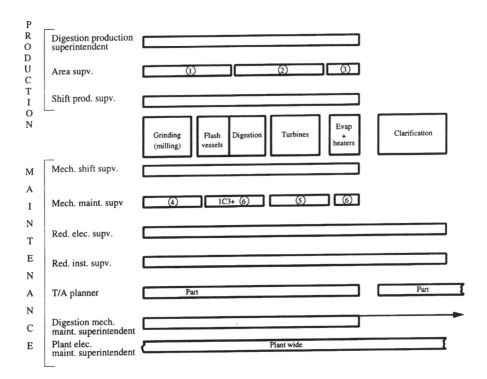

Figure 5.4   Comparison of supervisors' plant responsibilities, digestion process

Figure 5.3, which indicates the roles of all of those who either operated or maintained the digestion sub-process of the refinery, exemplifies the maintenance characteristics of the administration.

Mapping the organization chart is an essential part of the author's audit method. This information is supplemented with inventories of the personnel (see Table 4.2) and also various administrative models (see, for example, Figure 5.4).

Traditional views on administrative management, and some guidelines

A link between levels, like that shown in Figure 5.3 between the Digestion Mechanical Superintendent and the Mechanical Maintenance Supervisor (Grinding), is the key manager–subordinate relationship. The essence of this is that the supervisor has the *responsibility* for ensuring that his own and his team's work achieve the desired results. For this, the supervisor must have the *line authority* over decisions within his responsibility area. The superintendent *delegates* duties (in this case   mechanical maintenance) to the supervisor and also the authority for the supervisor to use the necessary resources (in this case trade group (c), see also Figure 5.3). The supervisor is *accountable* to the superintendent for achieving the desired results. The superintendent remains accountable for this work to the maintenance manager, i.e. authority is delegated as far down the line as possible but responsibility is not shed by doing this (see Figure 5.5).

One man can only effectively manage a limited number of subordinates. It has been suggested that this number lies somewhere between three and twelve depending on the complexity of the decision making (the Digestion Mechanical Superintendent, for example, has a *span of control* of five). Because of this constraint most organizations comprise several subordinate management levels. In the example of Figure 5.3 the chain of command passes down through five levels, from managing director to shop floor. Because there are clear advantages in having a short chain of command, some compromise must be reached between the length of this chain and the span of control (sometimes achieved by breaking a large structure into smaller ones – e.g. *manufacturing units*, see later – a form of decentralization).

The foregoing observations refer only to line relationships (the transmission of decision-making power down through the hierarchy and into the various work areas) but the 'horizontal' transmission of information and, in certain cases, of decision-making power is also needed. The principal relationships here are:

(i)   *Collateral* – those in which the work carried out in one area impinges on that of another, independent, area (i.e. between the Shift Mechanical Supervisor and the Shift Production Supervisor of Figure 5.3).

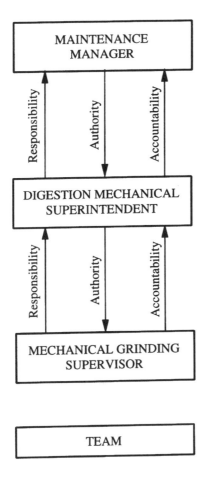

Figure 5.5   Formal relationships in the administrative structure, digestion process

(ii)   *Staff* – those in which the occupant of a managerial role has staff
        authority in a defined area of another manager but does not have
        line (managerial) authority over him. In the example of Figure 5.3
        the Utilities Manager gives the Digestion Mechanical Engineers the
        authority to make decisions and to give instructions to his staff in
        the limited area of design-out maintenance.

Classical theory dictates that each individual should be responsible directly
to one person only; this principle – of *unity of command* – is the basis of the line
relationship. However, a number of modern structural arrangements modify
this in dividing an individual's activities so that he is responsible for different
duties (or aspects of the duties) to different managers. Such a structure would
occur in Figure 5.3 if a Digestion Area Group (all personnel within the dotted
line) were to be formed, with group objectives and responsibilities and a

group leader. Staff within the group would then report both to the group leader (for all work carried out in the digestion area) and to their functional manager. Where there is more than one such plant-operating group in a large process plant the organization can be called a *matrix*. In order to avoid conflict in such arrangements the two managers must communicate closely about the duties of their subordinates and about the way they convey their instructions.

Even small organizations can have many complex relationships and it is therefore advisable that there is a job description for the work of each individual in the hierarchy. This should set out in clear, unambiguous, terms the job's

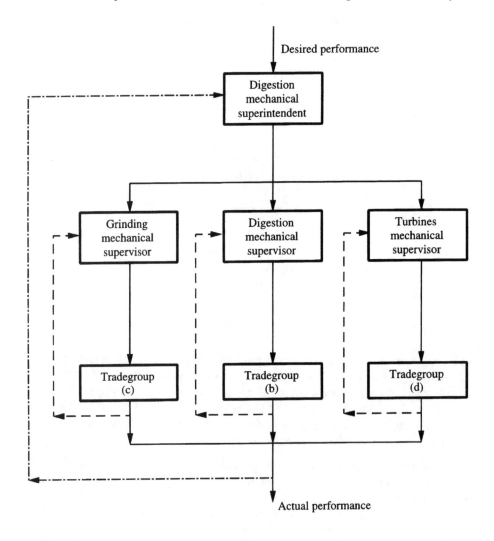

Figure 5.6   Administrative control system, digestion process

main functions and objectives, and the individual's limits of responsibility and authority, both financial and with regard to personnel. It should state to whom and for whom the individual is responsible; this should include staff and other dual reporting relationships.

Because of the inter-disciplinary nature of most maintenance work it is also necessary to:

(i)   establish standing committees for joint decision-making areas (in the situation of Figure 5.3 there is a weekly meeting for maintenance work-planning, involving the maintenance and production supervisors, a stores representative and the maintenance planner);

(ii) establish *ad hoc* committees for special projects.

To conclude this survey of traditional administrative theory it is necessary to say something about *administrative control*. The basic system by means of which a manager controls his team is illustrated in Figure 5.6. The Digestion Mechanical Superintendent is concerned with work and with decisions which involve a time scale much longer than that of the decisions of his supervisors. The former carries out his task by communicating the necessary instructions, and the aim of the work, to his supervisors. They, in turn, instruct their work groups on how to complete their tasks. Information feedback to the supervisors enables them to control the completion of the work in the short term; information feedback to the superintendent enables him to control the performance of his supervisors, and hence the completion of the work, in the long term. This is an example of a *vertical control system*.

Characteristics of maintenance administrative structures

Having reviewed some of the key points of classical administrative theory we are now in a better position to identify and discuss some of the principal difficulties of administrating the maintenance resources.

(1)   The maintenance–engineering interface

It is usual (see Figure 5.3) to separate the engineering responsibilities for the procurement of *new* plant from those for the maintenance of *existing* plant. Although this has advantages it also presents difficulties, mainly caused by lack of clarity in the overlapping areas of engineering responsibility for the plant.

Situations in the mining industry have been observed where the engineering department was responsible for mobile equipment when it was off-site (e.g. out for contract overhaul) and the maintenance department was responsible for the maintenance of the equipment when it was on-site. There was confusion over who 'owned' the equipment, who 'owned' the maintenance budget and who should specify the overhaul work (see Example 1 in Chapter 7).

The engineering department often 'owns' the professional engineers responsible for maintenance improvement (i.e. for design-out maintenance). The Digestion Mechanical Engineer of the Figure 5.3 example was located in the digestion plant area but reported to the Engineering Manager. Conflict existed regarding his job priorities – should he have been concentrating on project work in the digestion plant or on design-out maintenance? Design-out ended up being neglected in spite of the low reliability exhibited by the plant.

Major difficulties are also experienced in the feedback of maintenance information to aid the specification of new plant.

The major problem here is usually lack of adherence to the basic rules of administration. The overlapping areas of responsibility (arising from collateral relationships), for example, must be clearly defined and understood by all.

(2)   The maintenance–production interface

Conflict, can, and often does, occur across the maintenance–production interface. This again is mostly caused by lack of definition of overlapping responsibility areas. This can lead to the entrenched view that 'they (the operators) bust the plant; we (the maintainers) fix it'. This causes 'horizontal polarization' – a conflict of attitudes and communications between the various groups of an organization, see Figure 5.7. This is most evident in large organizations of the type shown in Figure 5.3, highly functionalized at the top with long chains of command down to the operators and maintainers – it is then very difficult to get the many disparate groups shown within

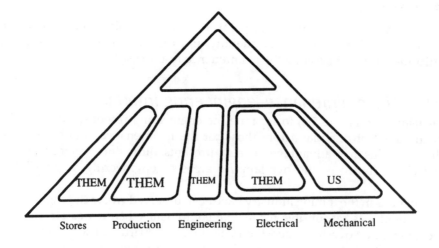

Figure 5.7   Horizontal polarization in an administrative structure

the dotted line of Figure 5.3 to work together to drive the plant.

The example of Figure 5.3 illustrates that large functionalized organizations with centralized authority for maintenance and engineering decision making have many advantages, including, *inter alia*

uniformity of engineering standards,
high level of technical knowledge,
high level of craft knowledge and skills,
easy introduction of new technologies on site.

These are particularly important when an extremely large and integrated process, such as a petroleum refinery, is being run. Centralized administrations, however, have difficulty in establishing 'ownership', a vital ingredient in the successful operation and maintenance of plant. In the example of Figure 5.3 this characteristic could be introduced via the following actions:

(i)   identifying process areas which have a clear production function and which can support a *plant operating group* (POG, see, for example, the Digestion group of Figure 5.3, shown within the dotted line);

(ii)   identifying the production and maintenance objectives for such groups;

(iii)   identifying the scope for inter-disciplinary teams (of up to ten personnel) within the group (The team objectives will need to be identified and made compatible with the group objectives). The team may also focus on a plant or process area. One way of identifying the teams is via 'equipment responsibility diagrams' of the kind shown in Figure 5.4. Such teams can become self-empowered;

(iv)   carrying out a major educational and training programme aimed at production–maintenance–engineering group and team building.

These changes may result in a modified administrative structure, as in the Figure 5.8 proposal where a Digestion operating group is suggested. In many respects, and especially if a group leader were to be appointed, this would be a matrix structure, i.e. the functional reporting structure would have been retained but the group members would also focus, as a group, to 'drive the plant'.

An alternative (see Figure 5.9) might be to create plant manufacturing units (PMUs) – sometimes called mutual recognition units (MRUs)[1]. This would differ from the matrix structure in that it would involve a structural change, the functional reporting set-up being divided into smaller units. The Digestion area would become one of a number of manufacturing units within the company. It would have considerable autonomy, having its own budget and management. Such an arrangement is more suitable where a company is made up of a number of clearly identifiable plants. Such units have not proved particularly successful for integrated plants.

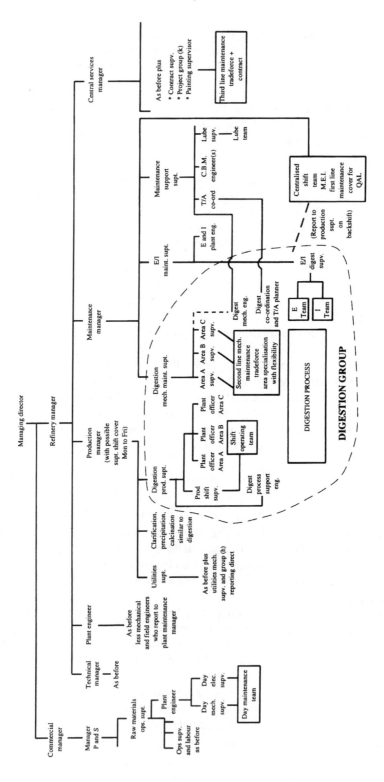

Figure 5.8   Proposed administrative structure based on plant operating group

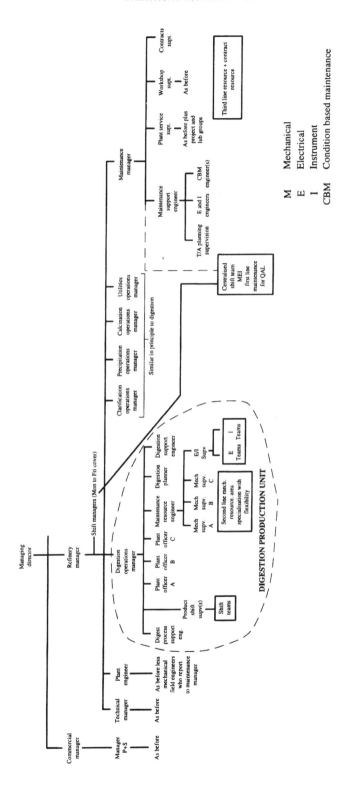

Figure 5.9 Proposed administrative structure based on production units

(3)   Responsibility for spare parts management

The rational objective for spares holding is to minimize the total of procurement, holding and stockout costs. Traditionally (see Figure 5.2), the responsibility for spares management has lain outside the maintenance domain. Thus, Maintenance specifies the spare parts, sets the *initial* order level and uses the parts, while the commercial department is responsible for the cost of the stores and the spares inventory policy. The natural tendency is for Maintenance to over-specify and overstock and for the commercial department to do the opposite. The responsibilities of the holder of spares and the user of spares must be clearly identified and systems of communication established for which the rules of operation (checks and balances) are clearly described and understood (see also Chapter 11).

(4)   Vertical polarization

Considerable antipathy can build up between the various levels of an organization – especially if they are numerous and the organization is large. (This has been particularly the case in the UK.) The greatest degree of antipathy is often between the shop floor and the higher levels of management (*vertical polarization*) – a conflict in attitudes, objectives and communication (see Figure 5.10). This can cause problems throughout the organization (not just in the maintenance area). For example, at shop floor level the characteristic that is of particular importance, and is diminished by such polarization, is *goodwill towards the company*. This is perhaps the most important of the human factors, probably the dominant one. The impact of other human factors, such as motivation and the sense of equipment ownership, stems from this.

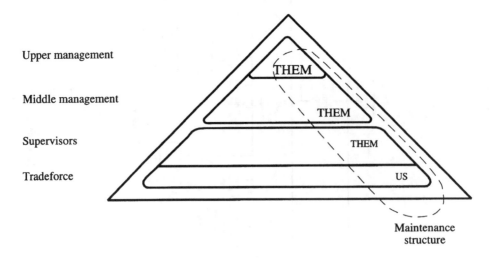

Figure 5.10   Vertical polarization in an administrative structure

Moves towards 'manufacturing units' or plant operating groups can have a positive effect in reducing vertical polarization. At shop floor level the introduction of plant-oriented teams and self-empowerment can help to break down antipathy towards management. The largest negative factor in this area, however, is lack of shop floor job security. This can have a negative effect on all the human factors.

(5) The relationship between the professional engineer and the maintenance supervisor

Supervisors mostly come from the trades, do not have professional engineering qualifications and only rarely move into the upper reaches of management. They are, however, unique in that they constitute the only level of management that looks downwards to non-management personnel. In addition, they tend to be less mobile within the organization than professional engineers and are the main source of trade and plant-oriented knowledge. More recently, their direct man-management role has been threatened by the implementation of self-empowered teams. This has been a principal cause of friction between supervisors and those above them.

A vehicle for achieving a better understanding of the above problem has been developed by Riddell[2] who considers that the traditional work roles of the maintenance supervisor can be represented by a grid of duties (see Figure 5.11) comprising four domains:

> Upward-facing Technical (UT)
> Downward-facing Technical (DT)
> Upward-facing Personnel (UP)
> Downward-facing Personnel (UD)

He points out that irrespective of the organizational changes that will take place these duties will always have to be carried out by someone. The trend towards the self-empowered team means that many of the UP and DP duties will be taken on board by the team and its leader. The UT and DT duties are tending to be carried out by technician-advisers and planners, who in general act in advisory positions to the teams. Clearly, these roles are key technical links between the professional engineers and the teams or the shop floor. As Riddell affirms, 'the supervisor has not become extinct, he has undergone a metamorphosis'. It is important that senior management recognize this and provide the necessary counselling and training.

(6) Major overhaul administration

A problem that is particular to maintenance management is the need, in many industries, to change the ongoing organization to cope with the demands of a major overhaul. Figure 5.2 showed the administrative structure for the

		Work diversity	
		Technical role	**Personnel role**
Janus - like traits	**UPWARD FACING** (Part of management team)	**UPWARD TECHNICAL (UT)** ● Influencing the maintenance goals ● Involvement in setting his own and his team's goals ● Influencing the maintenance strategy ● Involvement in setting preventive maintenance programme ● Involvement in work order and other maintenance information systems ● Using condition monitoring systems and equipment ● Collecting reliability and maintainability data and passing to engineers ● Advising on design-out maintenance ● Co-operating with other staff on technical/work matters in maintenance, production, stores, safety, engineering functions.	**UPWARD PERSONNEL (UP)** ● Communicating men's concerns and ideas - acting as their advocate ● Influencing personnel policies and decisions on - tradesmen and apprentice recruiting, selection, training, promotion, control and disciplinary procedures, pay differentials, bonus payments, overtime, amenities, dismissals, redundancies ● Influencing policies and decisions on - foremen selection, training, development ● Training new foremen and young engineers ● Giving advice on industrial relation problems and disputes, negotiations with unions ● Cooperating with other staff on personnel matters in personnel, safety, maintenance functions
	DOWNWARD FACING (Leader of own team)	**DOWNWARD TECHNICAL (DT)** ● Producing PM schedules in accordance with PM programme ● Making decisions on corrective maintenance - what is to be done, when, how and by whom ● Setting job methods and work standards ● Deciding on materials, tools, and information needed for each job ● Implementing maintenance systems and ensuring their continued proper use ● Montoring work output and performance, deciding corrective action to achieve team goals and implementing that action ● Investigating plant/equipment problems, seeking immediate solutions - if available implementing ● Developing and improving job methods, tools and work standards	**DOWNWARD PERSONNEL (DP)** ● Communicating the firm's department goals and policies Communicating team targets and plans ● Allocating jobs to men and maintaining team activity ● Motivating each member of team to achieve job targets ● Involving team in identifying new targets ● Guiding, training each man in (a) job knowledge and skills, (b) use of maintenance systems ● Setting behaviour and relationship standards, monitoring these in team and improving ● Controlling and disciplining men in accordance with agreed policies ● Resolving individual's personal problems ● Settling disputes and negotiating on minor industrial relations issues with shop stewards within agreed procedures ● Deciding on working conditions, hours, payments and amenities within agreed policies

Figure 5.11   A grid of the maintenance supervisor's roles

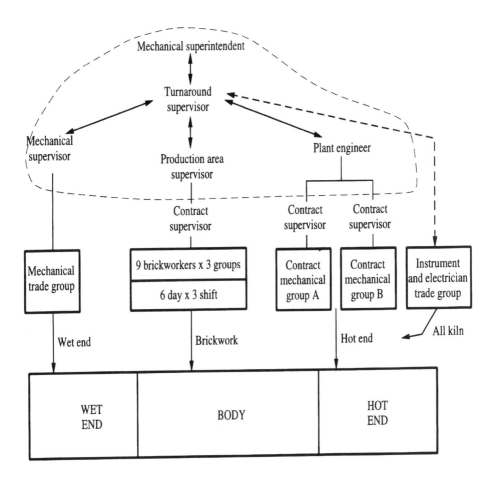

Figure 5.12 Administration for turnaround work

alumina refinery when it is in normal operation. When, however, major parts of the refinery (e.g. the kilns in the calcination area) are shut down for overhaul the resource and administrative structures in those areas have to change – to forms that are more appropriate for a *project* type of activity. The idea is illustrated in Figure 5.12, where it is indicated that the main link between the overhaul and the ongoing administrations is the Turnaround Supervisor, who also acts as Project Manager. This will be discussed in more detail in Chapter 9.

Summary

The problems discussed above can be divided into those that are general, i.e. items (4) and (5), and the rest, which are particular to the maintenance of plant. The basic difficulties highlighted in items (1) and (2) stem from the size of the operation. Large integrated plants make for large organizations and there is a pressure to specialize by function at a high level of the administration – to create a specialist engineering section which will set standards for equipment and also specify and procure it, to create a maintenance section which itself might be departmentalized by function into mechanical and electrical, etc. Thus, the duties and responsibilities for areas of the plant or of the process are usually set within each function, or trade, without thought for plant-operating groups and teams. See, for example, Figure 5.4 which shows a mis-match of the responsibility of supervisors across the plant process. Thus, in order to 'drive the plant' in such situations a major effort of co-ordination is needed.

When designing or modifying young organizations careful thought should be given to achieving the right balance between functionalization and the creation of plant-operating groups and teams.

The design or modification of the administrative structure

The design of a maintenance administrative structure is concerned with:

- determining the responsibility, authority and work role (the decision-making bounds) of each individual concerned directly with the management of maintenance resources;
- establishing the relationships, both vertical and horizontal, between each individual concerned directly or indirectly with the management of maintenance resources;
- ensuring that the maintenance objective has been interpreted for, and understood by, each individual concerned directly with the management of maintenance resources;
- establishing effective systems for co-ordination of – and for communication between – each individual concerned directly and indirectly with the management of maintenance resources.

Many of the rules and guidelines of classical administrative management – concerning such matters as chain and unity of command, span of control and so on – can be used to assist the design of a maintenance administrative structure. A procedure for such design, the aim of which should be to facilitate administration at least administrative cost, is shown in Figure 5.13.

Step (2) is concerned with the bottom-up influence on the *lower* structure – with the composition, duties and administration of the trade groups and first line tradesmen or tradesmen–operator teams. Steps (3) and (4) address the *upper* structure and take account of top-down influences, the former being concerned with the engineering needs of the structure – e.g. is there a need for a separate engineering group or can this be combined with the maintenance needs? – the latter with identifying the maintenance management requirements – e.g. should the managers be single discipline or multi-discipline? How many levels of management are required? Is a

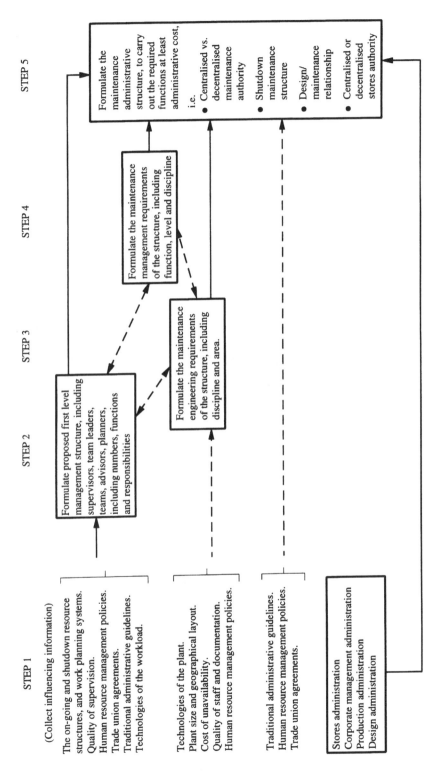

Figure 5.13   Decision procedure for formulating a maintenance administrative structure

separate planning section needed? etc. All of this has finally to be considered, Step (5), in the light of the rest of the structure.

During many years of consultancy the author has only once been involved in setting up an administrative structure for a new plant, the task has nearly always involved improving the structure for an existing plant and has been carried out as a part of an audit, after the life plans, workload and resource structure have also been reviewed. In such situations it is important to map and review the existing administration (of production and engineering) in order to identify the problems, which may be the result of having too many levels of management or of poor managerial performance. If such problems can be identified the objective of the review is to re-design the structure to enable it to better carry out its function and at reduced cost.

The mapping of the plant layout, and of the resource and administrative structures, of the alumina refinery (see Figures 4.1 to 4.3 and 5.1 to 5.4) were part of a maintenance audit. It can be seen from Figure 4.14 that a considerable immediate reduction in the level of the tradeforce numbers and cost was achievable – mostly by reducing the dependency on contract labour.

In the longer term, the development of improved inter-trade flexibility should lead to increased organizational efficiency and further reduced numbers. Clearly, this modified resource structure would have a considerable bottom-up influence on the existing administrative structure shown in Figures 5.1 to 5.3. For example, because the tradeforce would be reduced, and self-empowerment schemes implemented, fewer supervisors would be needed. In addition, there would be considerable top-down pressure to reduce the levels and concentration of management while encouraging production–maintenance integration. One way to accomplish this structural slimming would be to adopt the manufacturing unit approach of Figure 5.9.

References

1. Jacques, E. and Clement, S., *Executive leadership*, Blackwell Publishers, 1994, Oxford.
2. Riddell, H. S., 'A supervisory grid to understand the role of the foreman in the process industries', Proc Instn Mech Engrs: Part E, *The Journal of Process Engineering*, Vol. **203**, 1989.

6
Trends in maintenance organization

Introduction

During the course of the last twenty or thirty years considerable changes have been brought about in the way in which the management function in general, and the maintenance management function in particular, are organized. In this chapter these changes and ongoing trends will be reviewed, with the aim of identifying and discussing those of them which improve the efficiency or the effectiveness of maintenance organization. The models of resource structure and administrative structure that have been introduced in the previous two chapters will serve as useful vehicles for mapping the various developments.

Traditional maintenance organizations

Taken together, Figures 6.1(a) and (b) model what may be regarded as the customary maintenance organization of a medium or large size company in the 1960s and 1970s. Figure 6.1(a) shows that the first and second line maintenance groups would be plant-located and backed up by centralized specialist trades and workshops. Where necessary, the areas and the centre cascade work to contract labour. Figure 6.1(b), on the other hand shows that although most of the resources would be plant-located the authority for decision making would be centralized. Indeed, the upper structure would be highly functionalized; Production would be responsible for operating the plant, Maintenance for maintaining it. Engineering would be responsible for the design and procurement of new plant. Although there would be advantages in such an arrangement there would also be serious disadvantages (especially where the structure had become very large), namely:

- low utilization, because of the many small single-trade or single-shift maintenance and production groups – often manned up to the peak of a variable workload;
- vertical and horizontal polarization within the structure;
- high management cost due to an excess of hierarchical layers and functional positions.

In combination, these can result in poor organizational efficiency.
The main trends in the 1970s were aimed at improving shop floor

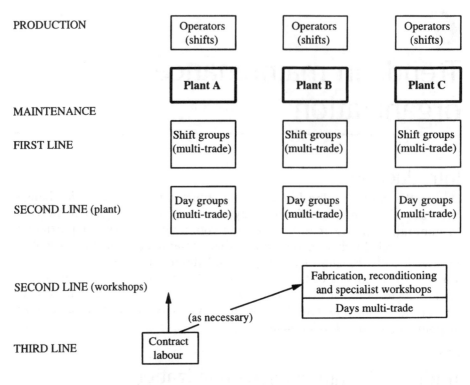

Figure 6.1(a)　Traditional maintenance resource structure

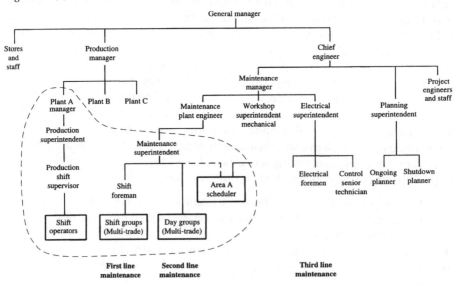

Figure 6.1(b)　Traditional maintenance administrative structure

utilization, this being achieved by increasingly centralizing the maintenance resource, see Figure 6.2(a). This was accompanied by improved work planning and stores control. In addition, extensive use was made of maintenance work measurement systems, such as Comparative Estimating (see Chapter 4, Reference 1), as a basis for tradeforce incentive schemes. These moves did improve tradeforce productivity. In the UK, however, the extent of such gains was limited by trade demarcation. There were many differentiated trades and many categories of non-trade workers; inter-trade flexibility was very limited and informal. In addition, centralization of the tradeforce and their supervisors had a negative effect in that it eliminated their feeling of plant ownership – at best they only 'owned' a job. It also exacerbated horizontal polarization.

Attempts were made to solve these problems administratively – by making the engineering staff responsible for the plant and locating them, as far as possible, next to the corresponding production staff (see Figure 6.2(b)). Technical officers reporting to the engineering staff were made responsible for sub-areas of the plant. The inspector-planners reporting to the technical officers were in turn responsible for still smaller sections of plant. *The engineering staff were the plant 'owners'.* Both the shift tradeforce and the

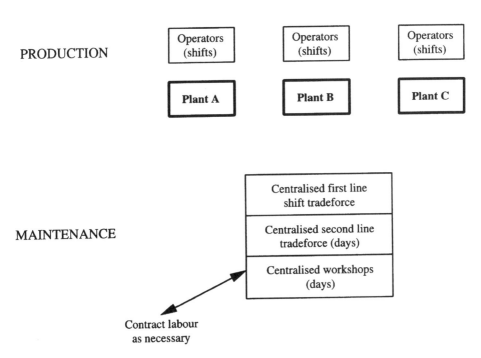

Figure 6.2(a)   Centralization of tradeforce with a view to improving its productivity

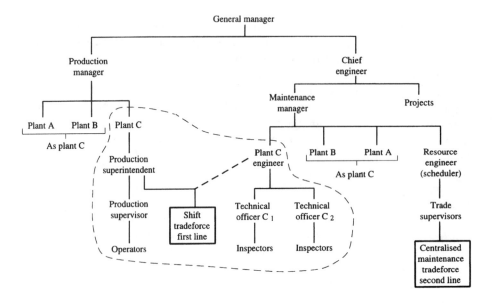

Figure 6.2(b)   Administrative structure – tradeforce 'owning' jobs, engineers 'owning' plant

second line resource were centralized, the former reporting to the shift production superintendent – reacting to plant-wide emergencies on a strict priority system – and the latter to a Resource Engineer via trade supervisors. Thus, second line jobs originating from plant areas would be planned and prioritized by inspector-planners and sent to the Resource Engineer for scheduling on the second line work programme. These moves clarified administrative responsibilities but did little to break down polarization or promote team working and true shop floor ownership.

The above approach was adopted at various plants of the Central Electricity Generating Board of England and Wales (a public monopoly, now broken up and privatized) and has also been used by several steel manufacturers, including BHP in Australia and Nissan Steel in Japan. Some petroleum refiners, such as Conoco, also went down this road but with local contractors being the maintenance tradeforce – even the resident second line groups (see Figure 6.3).

Broadly speaking, the productivity of tradeforces, whether centralized or not, steadily improved throughout the UK during the 1980s as a result of a sea change in industrial relations, the general evolution being as outlined in Figure 6.4 – although it progressed more rapidly in some sectors than in others (indeed, some industries are still at Stage 1 of Figure 6.4). In the late 1970s several companies, e.g. British Steel, negotiated productivity deals which reduced the level of unskilled maintenance labour while consolidating the numerous skilled trades into just three: mechanical, electrical and control. In the

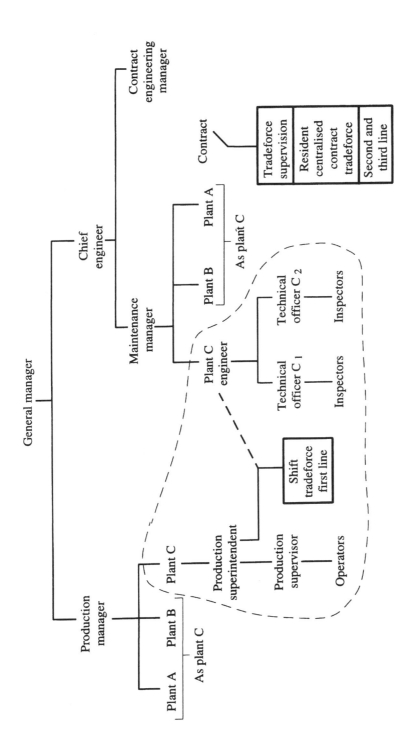

Figure 6.3 Administrative structure – contract tradeforce, engineers 'owning' plant

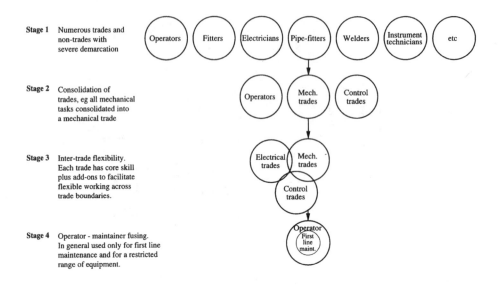

Figure 6.4   Trends in tradeforce flexibility

early 1980s a number of companies (also including British Steel) introduced cross-boundary skilling, a mechanical tradesman, for example, might be trained to work occasionally in an associated skill area – usually on specified equipment.

Although excellent examples of true multi-skilling can be found, its development is not as widespread in the UK as might be expected. The main reason for this is the high cost – in an environment of relatively high labour turnover – of the required training. In Australia a high-profile joint initiative of government and trade unions, financed in part by an industrial training levy, has promoted skill extension. This, however, appears to have been of more benefit to the tradesmen than their companies. The former have extended their skills, and in many cases their payment (as part of the agreement) and job mobility while the latter have not always achieved the skills profile that they needed – and in some cases have yet to decide how best to use the skills improvements that *have* been generated. Some Australian companies consider that skills improvement should grow out of company needs rather than from government and trade union initiatives.

Undoubtedly the most significant trend of the late 1980s and early 1990s has been the emergence of the operator-maintainer. Shell Chemicals, Carrington, were one of the first companies to promote this concept (this will be discussed in a later case study), the most successful application of which has been to first line shift work where – for a limited range of plant –

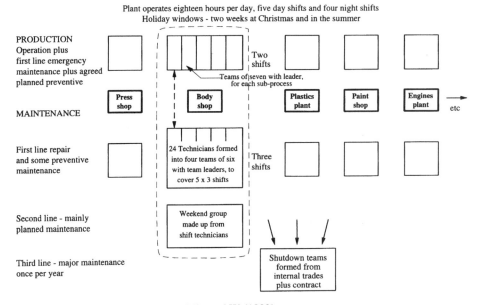

Figure 6.5(a)   Resource structure, Nissan UK (1993)

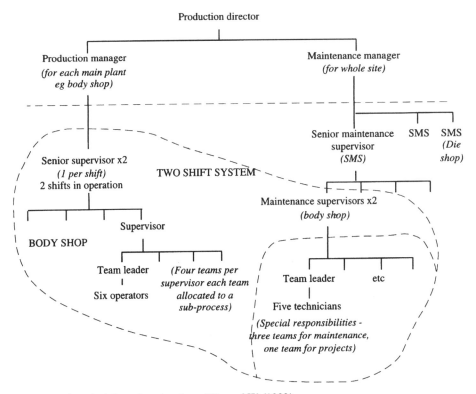

Figure 6.5(b) Administrative structure, Nissan UK (1993)

tradesman and operators have been trained both to operate that plant and to carry out first line maintenance across all the traditional trades. Where companies (including Shell Carrington) have tried to promote a similar approach to second line work it has not been so successful; the experience with this has been that the tradesman will always revert to his main (core) skill. To sustain the quality of his work he needs to practise his wider armoury of skills on a regular basis.

In the late 1980s a number of Japanese companies built manufacturing plants on green field sites in the UK and negotiated Japanese-type agreements with the trade unions – one factory union, no-strike undertakings, etc. – Nissan, in Sunderland, being a typical example. A noteworthy aspect of the latter development is that while retaining a traditional structure Nissan promoted the adoption of plant-oriented teams (see Figures 6.5(a) and (b)). In addition, they encouraged the sense of plant ownership by training their operators to carry out first line maintenance tasks and to be involved in a philosophy of continuous improvement, i.e. they fostered *autonomous maintenance teams*. What must be emphasized is the considerable level of training required to establish these – up to two years of effort is required.

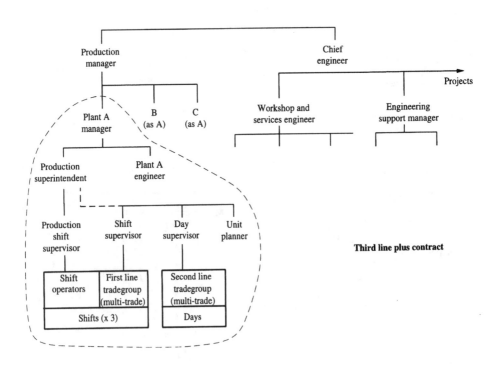

Figure 6.6   A PMU administration

In many respects, the Nissan structure of Figure 6.5 implements the idea of plant-operating teams that was put forward in Chapter 4.

Plant manufacturing units (PMUs)*

An alternative solution to the problems of very large structures is to divide them into smaller semi-autonomous units – the so-called 'ship' structure. This is a form of departmentation, e.g. the structure of Figure 6.1 is modified to that shown in Figure 6.6. In this case each 'manufacturing unit' is formed around a process (see Chapter 5).

The author first encountered this approach, and the human issues involved, in selecting and training PMU staff in British Steel in the late 1970s. In the 1980s and 1990s it seems to have developed into something of a fashionable trend, many companies having moved in this direction, including (in the UK) Shell Chemicals of Carrington and Courtaulds Chemicals of Derby, and (in Australia) Alcan and Queensland Alumina. Others, such as Glaxo in the UK, have used a similar approach but formed the unit around a product line. It

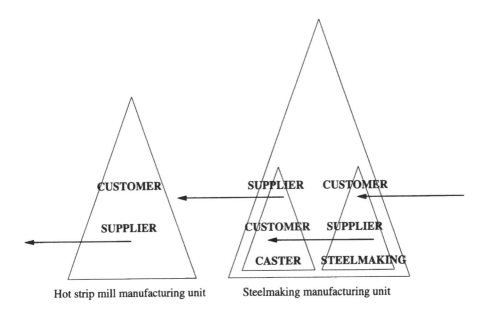

Figure 6.7 Practice of TQM across PMUs

*The difference between a *plant manufacturing unit* and a *business unit* is one of scale. The latter would be large enough to operate as a self-contained business (sales, production, maintenance, finance), but under the umbrella of a large organization. The former would be a part of a major plant but would have an identifiable product and would be semi-autonomous (production, maintenance, budget).

must be emphasized that such reorganizations usually come about for non-maintenance reasons and provide the following *general* advantages:

- Responsibility for costs and profits can be better identified.
- At PMU level, co-ordination and planning of work is improved.
- Setting of objectives and their translation into terms accepted by the staff and workforce are facilitated.
- The ideas and techniques of Total Quality Management (see Figure 6.7) are more easily introduced.
- *Esprit de corps* is improved.

The maintenance function also benefits from these general advantages. In particular there is easier co-ordination and execution of maintenance work. In addition, human factors aspects – such as shop floor goodwill towards the management, or an improved sense of equipment ownership – tend to improve.

The main general disadvantage in forming PMUs is the duplication of functions and equipment, which can result in a larger tradeforce. This is evident in many of the companies which have adopted the idea. In the example of Figure 6.6 it can be seen that the technical and engineering or maintenance expertise available within each unit is limited. This is a consequence of the dilution of centralized professional expertise and authority. It can result in each of the units going its own way in terms of maintenance strategy and engineering standards. The more successful examples of adopting PMUs (e.g. Alcan) guarded against this by clearly defining the responsibilities of the support teams and developing a clear understanding of the channels of communication between the support teams and the manufacturing units.

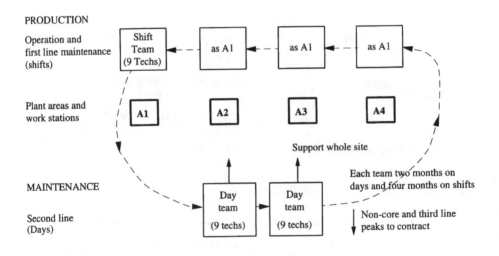

Figure 6.8(a)   Modernized resource structure, petrochemical plant

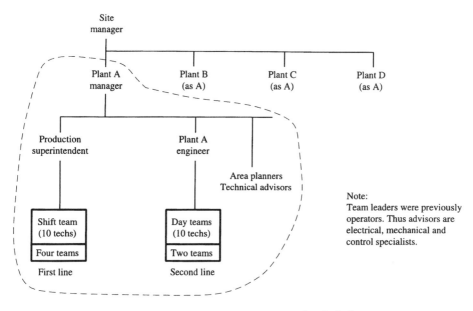

Figure 6.8(b)   Modernized administration structure, petrochemical plant

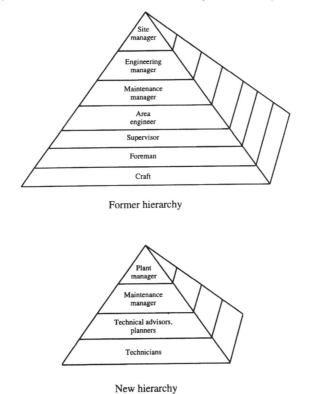

Figure 6.9   Reduction in number of levels of management, petrochemical plant

In addition, considerable importance was given to the use of an *Engineering and Maintenance Procedures Manual*. Such a document was used in all the individual PMUs and was regarded as an extension of centralized engineering authority.

Slimming the structure ('downsizing')

The company that exemplifies the principal maintenance organization developments of the last ten years is Shell Chemicals, Carrington, UK. The move from a large traditional structure into one divided into PMUs (see Figures 6.8 (a) and (b)) was accompanied by other significant changes:

- reduction in the number of management levels from six to three (see Figure 6.9);
- changing the role of the supervisor to that of a planner or technical adviser to self-empowered operator–maintenance teams;
- amalgamating operators and tradesmen into a single-role group, namely 'manufacturing technicians', who were assembled into process-orientated, shift-flexible, self-empowered teams – a single team working as operator–maintainers for four months followed by two months as second line maintainers. This was accompanied by extensive plant and process-orientated training. Each team had a leader whose core skill was in operations. The technical advisers were trade specialists;
- putting non-core work (such as reconditioning or building services),

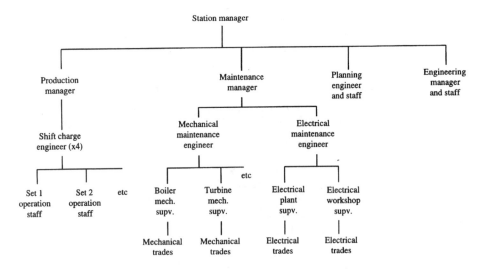

Figure 6.10   A traditional power station administration

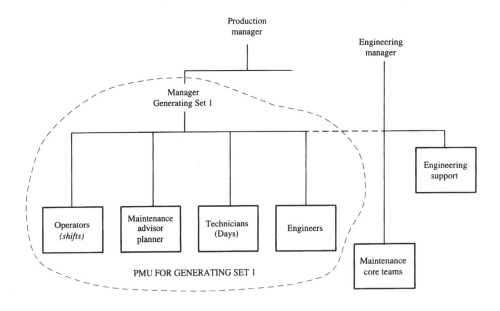

Figure 6.11(a)   Power station administrative structure based on responsibility by generating set

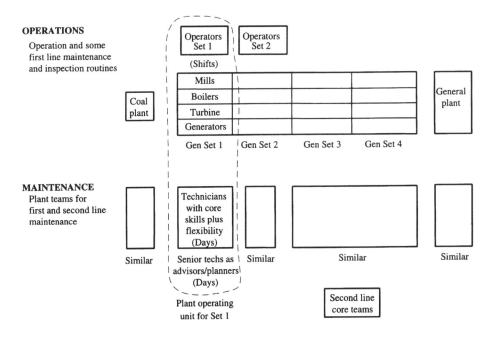

Figure 6.11(b)   Power station resource structure based on responsibility by generating set

and most of the major overhaul work, out to contract.

This re-structuring was implemented as revolutionary change in the mid-1980s but the structure, especially its flexibilities, has continued to evolve. In addition to the re-focusing of the operator and maintainer effort, major gains in labour and management productivity were made.

Many other companies in the UK and Europe have implemented one or more of the above changes – usually as part of an evolutionary development. An interesting example is to be found in the power generation sector. Here, the organization has customarily been as shown in Figure 6.10, Operation having responsibilities allocated *down* the process, i.e. by generating set, with Maintenance having responsibilities allocated *across*, i.e. by equipment type (boilers, mills, etc.). Several power stations have now set up structures of the kind shown in Figures 6.11(a) and (b); they have created PMUs by assigning maintenance workers down the process with the operations staff. In addition, considerable effort has been made to introduce self-empowered teams into these units.

The movement towards self-empowered process-oriented teams (SEPOTs)

The more advanced form of self-empowered team possesses the following characteristics:

- it consists of approximately ten shop floor personnel whose function is to operate and maintain a sub-process or area of plant (see

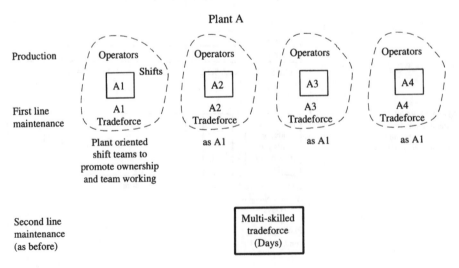

Figure 6.12   Plant-oriented teams

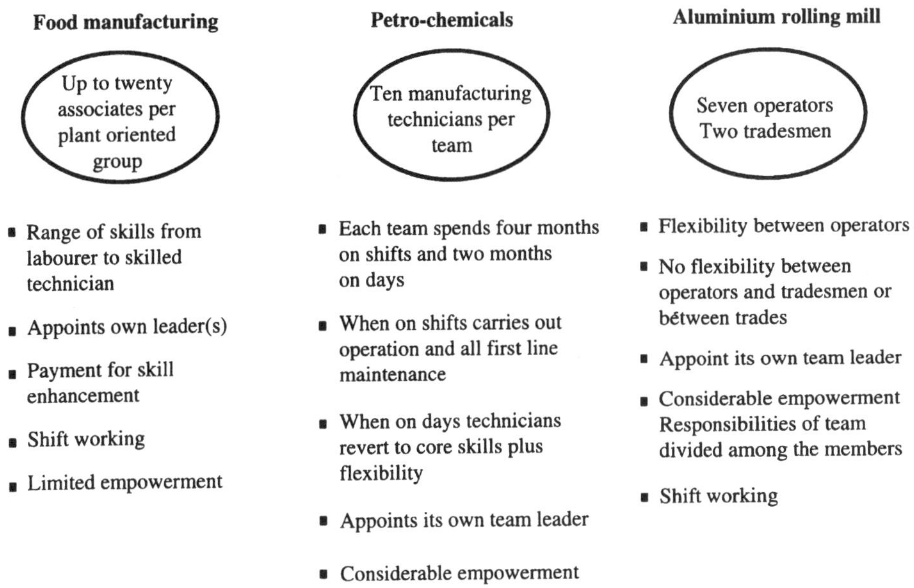

Food manufacturing	**Petro-chemicals**	**Aluminium rolling mill**
Up to twenty associates per plant oriented group	Ten manufacturing technicians per team	Seven operators Two tradesmen

Food manufacturing

- Range of skills from labourer to skilled technician

- Appoints own leader(s)

- Payment for skill enhancement

- Shift working

- Limited empowerment

Petro-chemicals

- Each team spends four months on shifts and two months on days

- When on shifts carries out operation and all first line maintenance

- When on days technicians revert to core skills plus flexibility

- Appoints its own team leader

- Considerable empowerment

Aluminium rolling mill

- Flexibility between operators

- No flexibility between operators and tradesmen or between trades

- Appoint its own team leader

- Considerable empowerment Responsibilities of team divided among the members

- Shift working

Figure 6.13   Alternative arrangements for plant-oriented teams

Figure 6.12);
- its members have various core skills but have been selected and trained to work flexibly in order to enable the team to tackle any job within its defined area of duty and responsibility;
- it has been empowered (allocated duties, responsibilities and authority) to carry out its day-to-day tasks with the minimum of *direct* supervision.

The teams at Shell Chemicals, described earlier, clearly have these characteristics as do the Nissan autonomous teams that were also discussed. In both cases the time needed for their introduction, not to mention the cost, was considerable – but so were the rewards.

In Europe the movement towards self-empowered teams has been carried mainly on the back of Total Productive Maintenance (TPM, see Book I, Chapter 14). Car manufacturers such as Rover (UK), Volvo (Belgium) and Renault (France) therefore feature strongly in its application. All have reported resultant large productivity gains.

Other than within the various TPM programmes, the introduction of SEPOTs has been limited. This may be partly because of cost and industrial relations difficulties but also because they are not always needed – in their advanced form at any rate. The size, type and characteristics of the teams

Figure 6.14   Star configuration of team responsibilities

will depend on the industry, the equipment and the workloads that they have to deal with (see Figure 6.13).

In food processing there is a high proportion of low-skill operator work and the composition of the plant operating team reflects this, containing many low-skilled operators and only a very few (or even one) highly skilled manufacturing technicians – Sagit of Naples designate the one manufacturing technician in each of their teams as the *conduttore*, or leader. Mars Confectionery, Slough have a similar arrangement. In both cases considerable effort has gone into training schemes, both to integrate the skills of the team and to increase flexibility between the traditional trades of the second line maintainer. Mars have introduced a motivational 'skill progression payment scheme', based on validated training and technical qualifications.

The aluminium rolling mill case of Figure 6.13 is an interesting example of SEPOT implementation. Although it was felt that inter-trade and operator–maintainer flexibilities were not essential, process-oriented teams were needed. Each shift team operated without a supervisor but with a technical adviser on day shift. The normal duties of a supervisor were divided among the team members (see also Chapter 7, Exercise 2).

The idea of self-empowerment has not only been applied to process-oriented teams. In the aluminium rolling mill case self-empowered teams, called *core* teams, were also developed to carry out the second line maintenance.

	1965	1975	1985	1995
Trends in resource structure	Decentralised	Some centralisation	Decentralised	Decentralised
	Multi-trade	Multi-trade	Inter-trade flexibility	Inter-trade flexibility
	Numerous non-trades	Numerous non-trades	Reduction of non-trades and number of trades via consolidation	Production - maintenance flexibility of first line work
	Strict demarcation	Strict demarcation		
	Poor resource flexibility ⇨	Poor resource flexibility ⇨	Improved resource flexibility including shift, inter-plant and use of contract labour ⇨	Increased resource flexibility, including * Non-core to contract * Annualised hours * Inter-plant and cascade
		Improved work planning systems, including work measurement		
Trends in administrative structure	Large structure	Large structure	Structured into business units	Structured into business units
	Highly functionalised	Highly functionalised	Still functionalised	Use of plant operating teams and other empowered teams
	Maintenance and production separated up to General Manager level	Maintenance and production separated up to General Manager level	Maintenance and production separated up to Business Unit Manager level	Maintenance and production joined at tradesmen level for first line work
	Centralised authority for maintenance	Centralised authority for maintenance	Introduction of plant operating teams	Extensive slimming of structure
	Single trade supervision	Single trade supervision	Multi-trade supervisors	Supervisor role changes to advisor/planner
	Large chains of command	Engineering and/or maintenance ownership of equipment	Influence of TQM and TPM	Extensive influence of TQM and TPM

Figure 6.15 General trends in maintenance organization

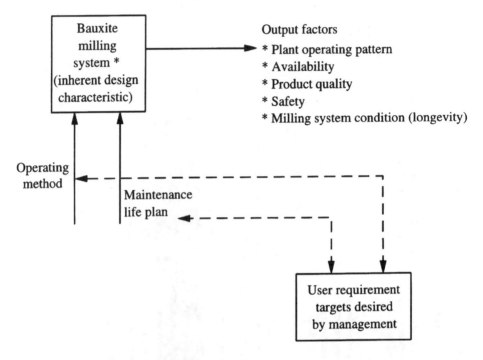

* Major part of digestion process - see Figure 5.4

Figure 6.16   The influence of operating method and life plan on output

The SEPOTs carried out the first line maintenance on shifts but the core teams carried out the major repairs and preventive work on the weekly downday – the SEPOTs owned the plant, the core teams owned the jobs. The core teams were given their duties and made their own decisions according to, and within, pre-determined guidelines. Administrative tasks (concerning safety, co-ordination, etc.) were allocated on a rotating basis, six months at a time per team member (see Figure 6.14). This concept is known as 'star' team self-tracking and regulation.

Summary
The trends in maintenance organization that have been outlined are summarized in Figure 6.15. The most influential ones have been:

(i)   The great improvement in tradeforce flexibility, between trades, between production and maintenance, between areas of plant, in shift working and in the use of contract or temporary labour.

(ii) The gradual merging of the production and maintenance functions – moving from the situation in the 1970s, where there was total separation at company level, to that of the present day, where they are merged at plant unit level.

These trends are complementary; without improved flexibility it would be impossible to achieve true self-empowered plant-oriented teams. The move towards such teams has been, furthermore, the most important single trend as regards the administrative structure. This can be explained by reference to Figure 6.16. If the desired performance (product output, product quality, safety and plant condition) is to be achieved from the bauxite mill it is essential that the operators and maintainers work closely together at that level – they cannot get any closer.

The Japanese recognized the need for this kind of amalgamation at a very early stage of their industrial regeneration (it can be seen that in the Nissan organization outlined in Figure 6.5 there is extensive use of self-empowered process-oriented groups). Their view, probably correct, is that it does not matter greatly whether the structure is large or small as long as you get it right at the bottom.

7
Exercises in maintenance organization

Introduction

Four exercises will now be presented that will afford the reader the opportunity to test his knowledge of the organizational principles that have been developed. *Solutions are given at the end of the chapter.*

Exercise 1   Restructuring the maintenance organization of a coal mining operation

Background

Figure 7.1 shows that 'COALCOM' comprises three underground collieries – operating three shifts per day, five days per week, fifty weeks a year – and a coal preparation plant. The coal is transported to the preparation plant by truck. In the short term, the surface coal storage isolates the colliery supply from the rail demand.

COALCOM organization

At senior management level the administrative structure is as shown in Figure 7.2. At this level the collieries and preparation plant function as plant manufacturing units. An engineering manager (with a secretary) has recently been appointed to assist in the co-ordination of the decentralized engineering departments, which carry out capital project work and have responsibility for the off-site overhauls of major equipment that is shared between the collieries.

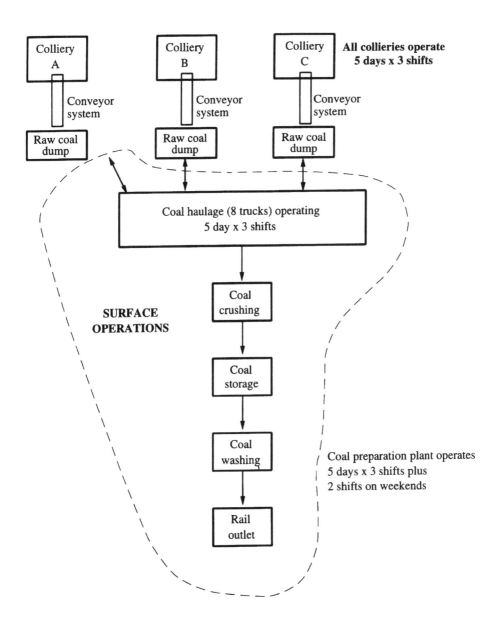

Figure 7.1 COALCOM process flow

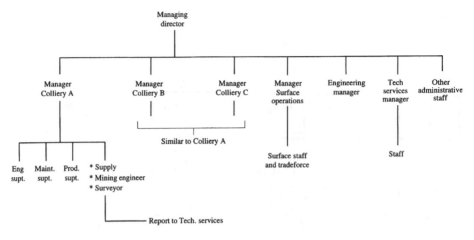

Figure 7.2   COALCOM senior management administrative structure

Colliery 'A' organization

Because all three collieries operate in a similar way, colliery 'A' can be used to illustrate COALCOM's maintenance strategy and organizational philosophy. Its layout is shown in Figure 7.3. Continuous miners* are used to develop the production areas and the tunnels for conveyor or worker access. Coal extraction (from the production areas) is achieved by 'longwall' cutting.** Clearly, a balance must be maintained between the rates of development and of production.

In summary, the underground plant comprises continuous miners, longwall equipment, coal conveyors and diesel-driven vehicles (such as the trucks for transporting workers). The maintenance life plan for each unit is made up of services and minor work carried out underground (lubrication, correcting malfunctions, etc.) and major work carried out on contract off-site (overhauls). The former is the responsibility of the maintenance department and is carried out mainly during two scheduled mid-week down-shifts, the latter of the engineering section. Figure 7.3 also indicates the locations of the tradegroups – e.g. '(h)' identifies the conveyor trade group; this coding will be continued throughout the various organizational diagrams occurring later.

Figure 7.4 shows the Monday to Friday maintenance resource structure. Only a small amount of labour was available for weekend work. The tradeforce is plant-specialized to provide a first line maintenance shift cover (e.g. the

---

* Diesel-driven vehicles with a front-mounted driller-cutter for cutting the development tunnels through the coal measures.
** Coal-extraction via a system comprising a shearer, armoured-face conveyor (up to 100 metres long), main conveyors and services (e.g. electricity supply). The shearer cuts slices of the coal seam (2 metres thick) by moving across a 100 metre block which has been developed between two tunnels. The removed coal falls on to the armour-plated conveyor and outwards to the conventional conveyors.

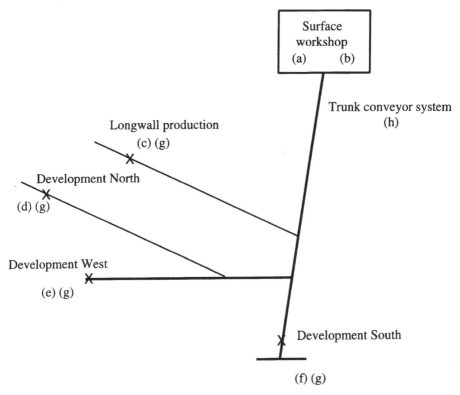

Figure 7.3 Layout of colliery 'A', showing operating areas

longwall has two fitters and an electrician – group (c) – on each shift). A small centralized pool, working in conjunction with the surface resource, moves to supplement the first line teams during mid-weekly downshifts.

During major underground maintenance work (mainly locating the longwall to a new production area) the resource is centralized as a shutdown group – little or no contract labour is used underground. The existing colliery 'A' administrative structure is shown in Figure 7.5, which should be looked at in conjunction with Figures 7.3 and 7.4.

A recently completed audit of these maintenance systems described the administrative structure as follows:

> The colliery manager is responsible for the process, equipment and manpower. Below this, administration is departmentalized at superintendent level into production, maintenance and engineering. It appears that the division of responsibility is as follows:
>
> *Production superintendent* – responsible for all aspects of the process and in this sense is the owner of the equipment while it is in operation;
> *Maintenance superintendent* – responsible for carrying out the maintenance of the equipment while it is on-site;
> *Engineering superintendent* – responsible for:

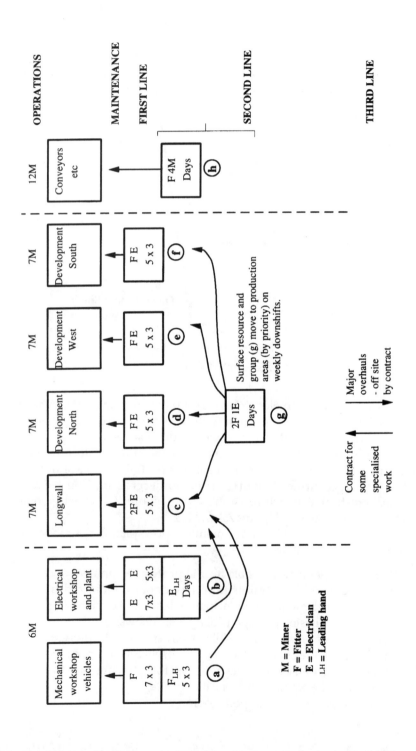

Figure 7.4   Colliery 'A' resource structure

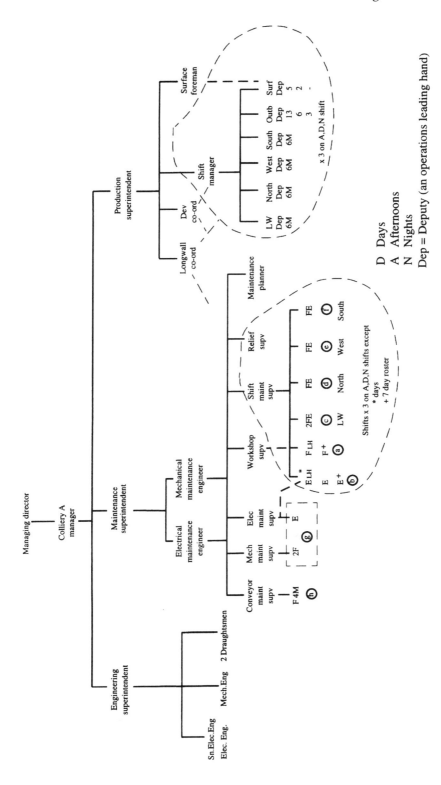

Figure 7.5 Colliery 'A' administrative structure

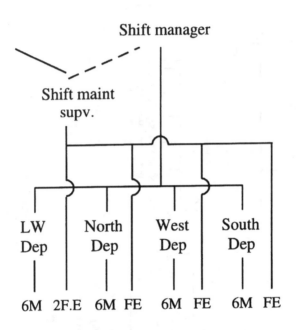

Figure 7.6   Colliery 'A' shift reporting

- specification and procurement of new plant and major modification work,
- specification and quality control of major overhaul work,
- engineering support to the maintenance department,
- maintenance of the equipment drawing and documentation systems.

The maintenance and engineering superintendents are mainly managers – they only get involved in 'technical matters' in a limited way. The main responsibility of the electrical and the mechanical supervisors is to organize the downdays and major maintenance in conjunction with the process co-ordinators. The maintenance planner is used as a maintenance documentation clerk.

The electrical and the mechanical maintenance engineers provide an additional layer of management. The former's function seems to be to support the 'management role' of the maintenance superintendent in the area of administration and industrial relations. ('I provide direct technical assistance for about ten per cent of my time.')

The main feature of the maintenance–production administration is the shift reporting structure – Figure 7.6 models this (and the production–maintenance relationships). At supervisor/shop floor level it has resulted in a number of process-related small production/maintenance teams.

The audit also made the following additional comments.

- In terms of satisfactory reliability and availability the most important activities are *the selection and procurement of equipment* and *the major maintenance work*. At COALCOM these are the responsibility of the engineering superintendents and are not done well.
- The first line shift tradeforce is poorly utilized.
- Relatively little progress has been made in the areas of inter-trade flexibility, operator maintenance and self-empowered teams. There is a considerable polarization (them and us) between the tradeforce and the management.
- The responsibilities of the engineering and the maintenance sections – for the maintenance of underground equipment – are not clearly enough defined.
- At superintendent level a considerable polarization of attitudes and perception makes communication difficult.
- The structure does not function as indicated in Figure 7.5. In practice, the electrical supervisor reports to the electrical maintenance engineer and the mechanical supervisor to his mechanical counterpart.
- There appears to be an 'organizational power base' at shift-manager level. Among other problems this was causing serious polarization between production and maintenance.

The problem

The management of COALCOM are concerned that the reliability of their underground equipment is low and maintenance and engineering costs high. They believe that the main problem is an inadequate organizational structure which they feel is in need of re-shaping

The question

The management have asked you to provide them with proposals for changing the engineering and maintenance organization. At this stage they only want an outline of the revised administrative structure, in the form of a list of the main changes that you feel are needed to improve the organization.

Exercise 2   A cautionary tale of organizational change

Background

The process flow of the aluminium rolling process concerned in this case is outlined in Figure 7.7. Four years ago the company decided to update the mill by major capital investment in sophisticated control systems to improve

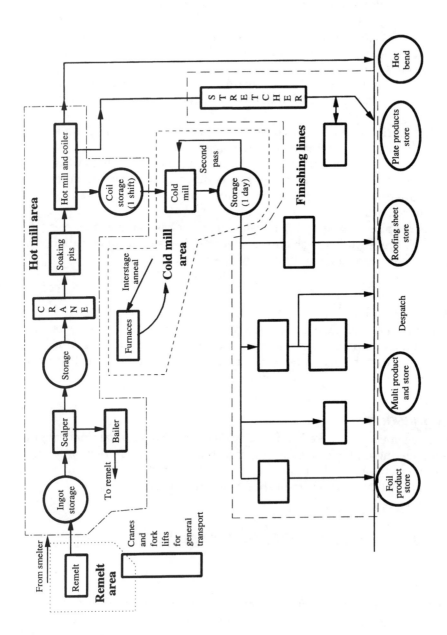

Figure 7.7   Process flow, aluminium rolling mill

throughput and quality of product. Some of the basic equipment was renewed and some (e.g. the hot mill) retained. The company realized that to operate and maintain the new plant they would have to restructure and invest heavily in improving the skills of the workforce. In addition, there was major pressure to integrate production and maintenance and to introduce the ideas of self-empowerment.

Before modernization the organization had been a traditional functional organization, i.e. there were many single-trade maintenance teams reporting to a centralized engineering manager. A large centralized trade group was responsible for the workshops, building fabric and services. In addition, there were area trade groups responsible for first and second line maintenance of the production plant. Figure 7.8 attempts to capture the essential characteristics of the organization after restructuring, which were as follows.

- Process-orientated groups (e.g. for the hot mill) were established, each under its own manager. Each had its own unit engineer and process engineer and in the case of the hot mill there was also a control technician to ensure that the group held the correct mix of engineering skills (the mill engineer was a mechanical). The shift teams each comprised six operators, an electrician and a fitter, were self empowered and operated a 'star configuration' of duties (see Figure 6.14). The function of the shift tradesman was to carry out first line maintenance. Each group had a degree of autonomy in terms of its production and maintenance policy.

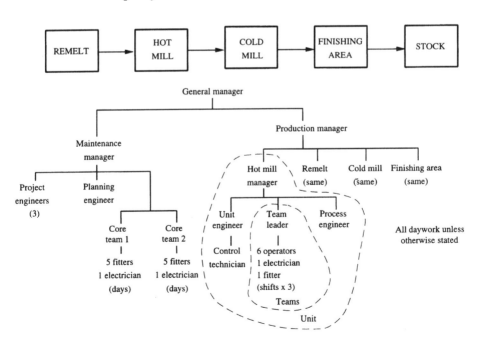

Figure 7.8 Rolling mill administration

- The process-orientated groups were supported by a centralized structure which included limited engineering support and two second line maintenance ('core') teams. Because each plant area could be scheduled separately – for one day a week – for maintenance, the core teams carried out all second line work with the help of the local tradesmen.

The problem

The main problem is the poor reliability of the hot mill area – mostly due to control and electronic systems reliability. However, there are some electrical and mechanical problems as well (which can arise when new control gear is mounted on old mechanical process plant).

There does not seem to be the expertise within the hot mill group to solve these problems. The hot mill engineer does not have the necessary knowledge in the control area. There also appears to be considerable difficulty in communication between the project engineers and the hot mill group, the former feeling that they should concentrate on project work. Although expertise is available it is held within the other process-orientated groups; communication with and between them is poor and each group has its own problems. In addition, morale within the hot mill teams was breaking down.

The questions

(a) Would you consider that the hot mill group could be defined as a 'business unit'. If not, how would you define one?

(b) Do you think that the hot mill shift team qualifies to be called a self-empowered team? What are the essential requirements for this?

(c) What would be the initial organizational action(s) you would take to help overcome the reliability problems of the hot mill?

(d) What organizational and/or procedural changes would you make to ensure that, in future, the necessary engineering and technical expertise could be focused on problem areas of plant.

Exercise 3   The changing role of the maintenance foreman*

Background

This exercise will involve the concepts and principles incorporated in Riddell's work-role grid – shown in Figure 5.11 – which categorizes the duties and responsibilities of the traditional foreman.

* Contributed by Dr H. S. Riddell, Honorary Fellow, University of Manchester School of Engineering.

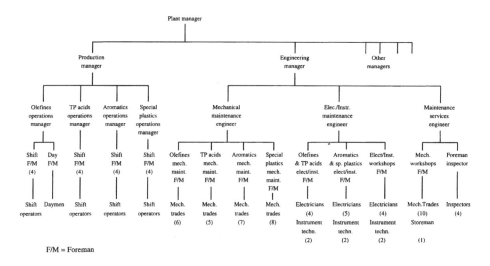

Figure 7.9 Administrative structure, petrochemical company

Part A: The foreman's role in a traditional organization

The administrative structure of a petrochemical company is shown in Figure 7.9. The operating policy of the company is to run its plant continuously, at its rated output, for three shifts per day, 330 days per year, the plant being shut down each August for a five-week major overhaul. The maintenance foremen and their teams carry out first line and second line work within their respective trades and areas. Emergencies arising outside normal daywork hours are covered by a call-out system for each trade and area. The plant teams are supported by the workshop for minor reconditioning and fabrication and by contract labour for work overloads and during the annual shutdown.

All of the foremen have at least six years' experience, have been promoted from the tradesmen's ranks and have been well trained in supervisory duties. In general, they are respected by both tradeforce and management. Their ages have a wide spread, with a mean of about forty-five.

Tasks and questions

(a)   For a foreman in the plant described, develop what you consider to be a full set of the duties and responsibilities falling in each of the categories UT, UP, DT, DP defined in Figure 5.11. Your approach should be based on your own experience and on Figure 5.11. Your answer should recognize that the foreman has no clerical, planning or technical assistance.

(b)   Make an estimate of the proportion of time that the foreman is likely to spend on each of the duties in each category. From this, estimate

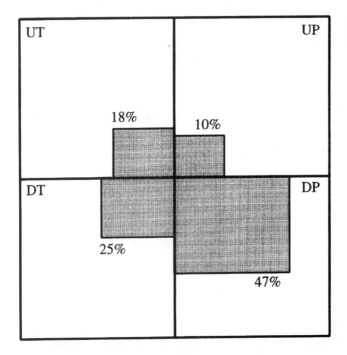

Figure 7.10   Proportion of maintenance foreman's effort spent in each category of duties

      the total proportion of time spent in each category and draw a
grid of the type shown in Figure 7.10.

(c)   Do your lists and grid show a balance or imbalance in the foreman's
range of duties and in the time allocation between working as a
junior member of the management team (UT + UP) and as a leader
of his own team (DT + DP)?

(d)   Do your lists and grid show a balance or imbalance in his range of
duties and in the time allocation between being responsible for
technical and plant matters (UT + DT) and being responsible for
personnel matters (UP + DP)?

(e)   How different are the personal behavioural traits needed by the
foreman to successfully carry out 'upward-facing' duties from those
needed for 'downward-facing' ones?

Part B: The role of the foreman after a restructuring exercise

After a restructuring exercise the administration outlined in Figure 7.9 was
changed to that outlined in Figure 7.11 where it can be seen that the levels
of management have been reduced (by the removal of the engineers and
operations managers) and the spans of control – in particular that of the

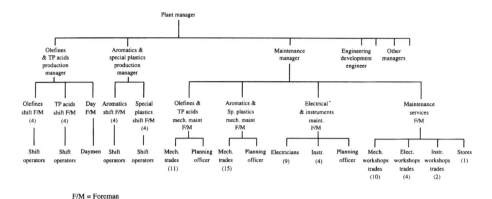

Figure 7.11   First organizational change in the administrative structure, petrochemical company

foremen – correspondingly increased. The foreman's increased span has been balanced by the introduction of planning officers to provide him with planning, scheduling and clerical support. The technical and plant knowledge of the planning officers is no greater than that of the foreman.

Tasks and questions

(f)   Re-examine your initial categorized list, question (a), in the light of the organizational changes outlined in Figure 7.11. Revise your time estimates and draw a new grid of the type shown in Figure 7.10.

(g)   To what extent do you consider that the increase in the foreman's DP and UP duties (resulting from his increased span of control) is likely to be compensated for by a decrease in his DT – and possibly UT – duties as a result of the support he now has from the planning officer?

(h)   What other changes could be made in the maintenance organization to support the foreman in coping with management's drive for 'flatter' structures?

Part C: Introduction of self-empowered work teams

Management are now considering introducing a programme of change leading to the establishment of self-empowered teams (SETs) involving both operators and maintenance tradesmen. A SET, operating in a maintenance situation, can be defined as:

> a group of employees who have day-to-day responsibility for managing themselves and the work they do with a minimum of direct supervision.

Members of SETs typically handle job assignments, plan and schedule work, make production and maintenance related decisions and take action on problems. To be most effective they need a wide range of skills, shared in a flexible manner within the team.

Tasks and questions

(i)   1. Identify, out of your question (a) list, the first batch of UT, UP, DT and DP duties which you consider should be transferred to the SETs (e.g. those needing shortest training, which are least critical, where quickest success is likely).
2. Identify those of the remaining duties in each of the four categories which, in your opinion, should be transferred and suggest a time scale for the transfer.
3. Are there any duties left which are still needed but which you do not consider appropriate to transfer to SETs?

(j)   Identify the changes or developments that are needed in the following positions once movement is made towards the introduction of SETs:

- maintenance manager
- maintenance foreman
- tradesman.

(k)   What part should the maintenance foreman take in the initial introduction of SETs?

(l)   What role (if any) do you see for the maintenance foreman after the SETs are fully implemented?

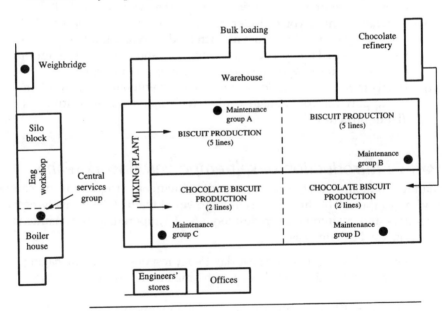

Figure 7.12   Layout of food processing plant, showing location of maintenance trade groups

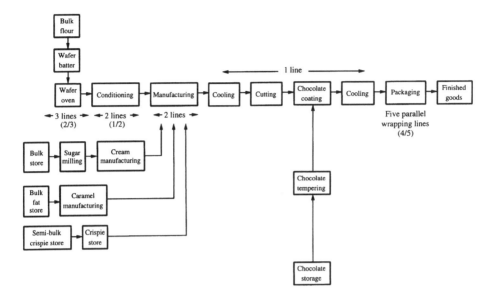

Figure 7.13   Process flow, chocolate biscuit line

Exercise 4   Maintenance reorganization in a food processing plant

Background

A food processing plant occupies a site with an area of some 50 000 square metres. It comprises ten biscuit-making lines, four chocolate lines, a chocolate refinery, mixing and preparation plant and general plant services (see Figure 7.12). The maintenance establishment numbers 95, twenty-one of whom are staff.

To a large extent, each production line is independent – although there are some common services at the front end of the whole plant. The chocolate-making lines, a typical one of which is shown in Figure 7.13, are far more sophisticated than those for making biscuits. Each of these lines (and its biscuit lines) makes a different product and they are not interchangeable. In general, they are made up of units in series (as are the raw material input streams, such as the caramel line), the failure of any one unit closing down the line. The chocolate lines, however, have some spare capacity (e.g., as shown, only two out of the three wafer lines are needed to keep the main line in full production). The plants operate three shifts per day, five days per week, fifty weeks per year. The weekends are used for production cleaning and maintenance.

The major maintenance is carried out during the annual fortnight's shutdown in the summer. In general, all the lines operate throughout the day and evening shifts, and about thirty per cent of them throughout the night shift also.

Company organization and maintenance strategy

The location of the maintenance 'plant groups' is shown in Figure 7.12, and the maintenance resource structure in Figure 7.14. The tradeforce is divided into five semi-autonomous groups, each carrying out the total maintenance workload for a designated number of lines (called plant sections) or plant services, i.e.

Group A :   five biscuit lines (Section A)
Group B :   two biscuit lines (Section B)
Group C :   two chocolate lines (Section C)
Group D :   two chocolate lines (Section D)
Group E :   the common services.

Each group carries out first line maintenance during the weekday production shifts, the second line work at weekends, approximately half of each group coming in at the weekend for the latter purpose. The third line work is carried out during the annual shutdowns, when each group is supplemented by contract labour. Each group remains plant-dedicated and to a large extent operates autonomously; there is only very limited movement between the groups. It should also be noted that the 'wrapping fitters' work only in the wrapping area of each plant section, reporting to the section supervisors for industrial relations purposes but receiving technical advice as necessary from 'wrapping supervisors'. There is little or no inter-trade flexibility and none at all across the operator–maintainer divide, strict trade demarcation rules applying within the company, which has a poor history of operator and tradeforce training. There are three main engineering trade unions represented.

In addition to the above plant-dedicated groups there is a centralized night shift group that carries out first line maintenance (overspill from the day and evening shifts) plus some inspection routines and other minor planned work.

The *company* administrative structure, shown in Figure 7.15, indicates that at senior management level there is a traditional division of responsibility by product, with complementary responsibilities as regards production and maintenance, e.g. Section D production superintendent and Section D maintenance superintendent are both responsible for the same five biscuit lines – they are also located in adjacent offices in that plant area.

The *maintenance* administrative structure is shown in Figure 7.16 and should be looked at in conjunction with Figures 7.14 and 7.15. For each of the decentralized groups a basic work planning system for on-line and off-line work is used, although the majority of work from Monday to Friday is unplanned. The rather limited planning and scheduling for the weekend and shutdown work is carried out by the section engineers.

There is no maintenance control, i.e. no plant failure or repair information is collected, stored or analysed. Most documentation is carried out using hard paper systems, with the exception of Section C where a simple, single-user, microcomputer-based preventive maintenance programme (with spares lists) is used.

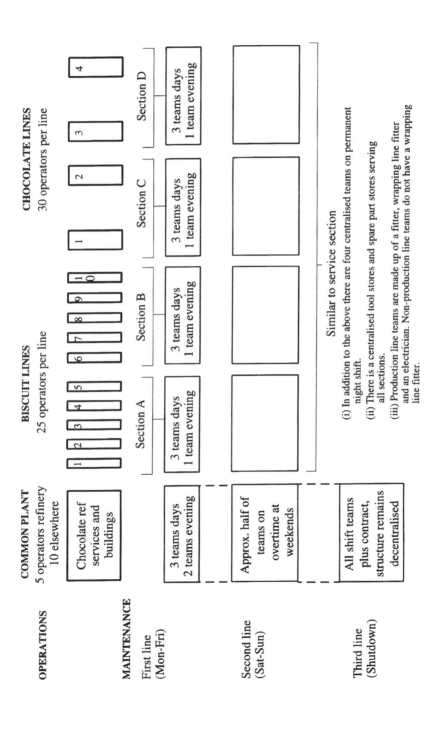

Figure 7.14 Resource structure, food processing plant

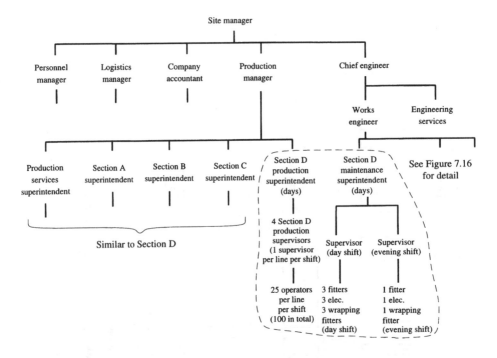

Figure 7.15 Company administrative structure, food processing plant

The type and quality of the maintenance life plan varies between the five groups, from a stated fixed-time inspection policy in Group C to operate-to-failure (plus lubrication) in Group A. Group C declared their approach to maintaining the plant to be as follows:

- Carry out an effective corrective policy, plus daily inspections, from Monday to Friday.
- Inspect at weekends and repair as necessary in order to keep the plant going throughout the following week.
- Carry out other fixed-time work at weekends, or during the annual plant shutdown, or by exploiting spare plant.

The problem

Senior management are concerned at the low level of production line availability and the associated high maintenance costs. In particular, they feel that the maintenance tradeforce is far from fully utilized during the weekly production shifts. Figure 7.17 shows the result of an activity sampling exercise, recently commissioned, which confirms this view (the corresponding graph for electricians shows a similar overstaffing). They also believe that the maintenance strategy is purely reactive and that even in Section C the stated preventive work is not being carried out at weekends because of the pressure of the unplanned corrective work. In addition, they are deeply concerned

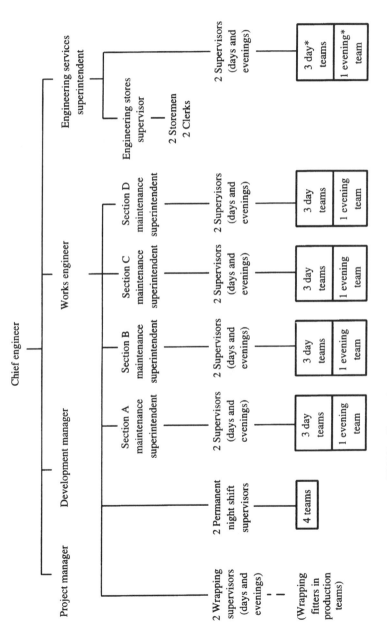

Figure 7.16   Maintenance administrative structure

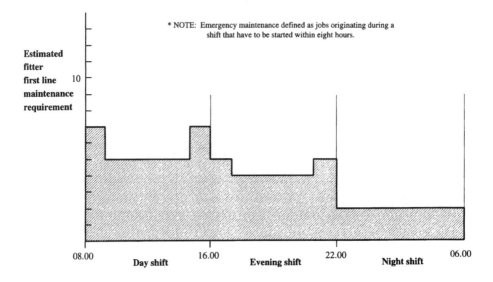

Figure 7.17   First line fitter workload

about the lack of co-operation between production and maintenance in each of the plant sections. They consider that there is no sense of plant ownership at shop floor level.

The questions

Senior management have asked you to consider the problem and to provide them with an approach for improving the situation.

Your answer should include the following:

(a) 1. A list of what you consider to be the main problem areas.
     2. The reasons why you consider the plant's condition to be poor and why it has gone out of control. An outline of the approach that should be adopted to remedy this during the next year.
(b) An outline of a modified organization which will facilitate the adoption of the approach you have outlined above. This should include the following:
     1. A proposed maintenance resource structure. This should take into consideration the possibility of more efficiently matching the first line shift resource to its workload (shown in Figure 7.17). It should also have regard for the need to carry out more planned work without increasing the size of the tradeforce (if anything it should be reduced).
     2. A proposed administrative structure, matching your proposed resource structure. Your presentation of this should show the main changes that you intend to make to both the maintenance and production departments.

Guideline answers

There are many possible solutions to each of the exercises and each of those given below should be regarded as just one of several ways of improving the maintenance organizational effort. The solutions offered attempt to use classical administrative theory in conjunction with state-of-the-art maintenance management ideas.

Exercise 1 COALCOM

The changed administrative structure is shown in Figure 7.18, the main changes being as follows:

- The formation of a centralized project engineering group reporting to the engineering manager, carrying out all project engineering duties and responsible for the off-site overhaul of equipment shared between the collieries.
- The removal of a layer of maintenance management from the colliery administration.
- The creation, in each colliery, of a maintenance support function which will provide direct professional engineering and planning help.
- Making the supervisors 'equipment dedicated' (e.g. 'Longwall Supervisor'), responsible for their own equipment, including the

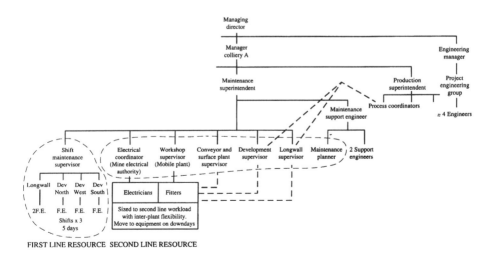

Figure 7.18   Proposed administrative structure, COALCOM

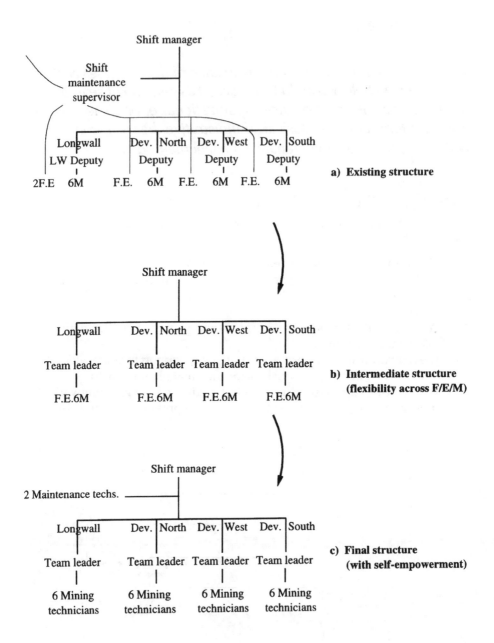

Figure 7.19 Moves towards self-empowerment, COALCOM

formulation and control of the life plans (and aided in this task by the support engineers).

- Arranging that the second line resource reports, in the case of electricians, to the Electrical Supervisor and, in the case of mechanical tradesmen, to the Workshop Supervisor – this group to be plant flexible and moving to the different equipment areas on downdays (e.g. when in the Longwall area, reporting on technical matters to the Longwall Supervisor).

- Envisaging that the first line shift resource will move through a number of stages, eventually operating as mining technicians, i.e. undertaking both production operations and first line maintenance (see Figure 7.19).

Exercise 2   A cautionary tale

(a) The hot mill group cannot be considered as a true 'business unit'. It is too small, is at too low a level in the organization and has only limited autonomy. A business unit would be responsible for a complete production process or product line – operating almost as a small company within a larger one. It would have autonomy over all its operations (i.e. including sales and finance) other than a relatively few specialized corporate functions. (An example of this would be a colliery within a mining company, or a block-making plant in an aluminium smelter.) The group is therefore best described as a small PMU (see Chapter 6).

(b) The hot mill shift team could be considered a 'self-empowered work team', which could be defined as follows:

A group of employees who have day-to-day responsibility for managing themselves and the work they do with a minimum of direct supervision. Typically, they handle job assignments, plan and schedule work, make production and/or maintenance related decisions, and take action on problems. To be most effective, they need to have a wide range of skills and to share the use of these skills in a flexible manner within the team.

(c) A group should be formed – from within the company – of engineers who have the necessary expertize (control/electrical/mechanical) in hot mill operation and maintenance. They should then be seconded, under the maintenance manager, to concentrate on the hot mill's reliability problems.

(d) It is essential that the unit engineers from each of the groups communicate with each other, via formal and regular meetings, to share their knowledge and to help each other. A limited degree of engineering support, in the form of project engineers or contract specialists, should be available from the central engineering section. It may well be that the hot mill unit engineer should be moved to one of the other groups on an exchange basis. Alternatively, he should have further training in control engineering.

Exercise 3 The changing role of the maintenance foreman

Part A

(a)   The list should be comprehensive, in line with the grid concept, but not a straight copy of Figure 5.11.

(b)   The grid is based on the proportions of time that the foreman spends in each of the categories – the total must come to 100%.

(c)   In general, the grid will show that the UT + UP duties are much less time consuming than the DT + DP duties.

(d)   The grid will show that the balance between the UT + DT and UP + DP duties will depend on the foreman's position, i.e. in the case of the instrument foreman the centre of gravity of the grid will be towards the UT + DT tasks; in the case of the services foreman towards the UP + DP tasks.

(e)   If the foreman is to successfully carry out his upward-facing duties he needs to be integrated and compliant; to carry out his downward-facing duties he needs to be self-assertive, a leader, and needs to have the ability to initiate ideas.

Part B

(f & g) The list should show a reduction in the DT duties as a consequence of having a planning officer (the duties having been transferred). Although the list of DP and UP duties may remain about the same, the time allocated to these will show a large increase as a result of the increased span of control. The centre of gravity of the grid will therefore have moved towards the DP category of duties.

(h)   Provision of effective maintenance documentation and information systems.
Provision of technical back-up by engineering staff.
Ensuring that each immediate supervisor provides adequate direction and assistance.
Recruitment of competent tradesmen.
Provision of plant-specific training for the tradesmen.
Promotion of, and training for, inter-trade flexibility.

Part C

(i) 1. UT   Advising on design-out-maintenance problems.
Involvement in setting up the SET objectives.
Involvement in improving the maintenance information systems.

UP   Involvement in changing working conditions.

DT   Assisting in the revision of preventive maintenance procedures.
Assisting in the setting of standard job procedures.
Monitoring work output and performance levels.
Monitoring work quality and safety issues.

DP   Job allocation.
Motivating team to achieve targets and objectives.
Monitoring team members' progress and problems.

2. The following duties could be considered for transfer to the SETs after a period of about twelve months of SET operation.

UT   The use of condition monitoring equipment.
Assisting in the revision of maintenance schedules.

UP   Influencing personnel policies for tradesmen and apprentices.
Involvement in training procedures for the SET members.

DT   Establishing job methods and work standards.
Involvement in the setting of improved team targets.

3.   The following four examples of those duties could be considered as lying outside the boundaries of the SET's responsibilities.

UT   Co-operating with other departments on technical matters.

UP   Influencing personnel policy on pay, promotions and discipline.

DT   Organizing on-site contractors.

DP   Disciplining individuals in accordance with agreed procedures.

Most of the foreman's duties listing in Figure 5.11 (and in your own listing) will continue to be relevant and will have to be carried out either by the SET or by some other member of the management. The exceptions will be some of the listed DP duties such as disciplining and controlling the men or guiding each man in job knowledge and skills.

(j) *Maintenance Manager*
The clarification of business and departmental maintenance

objectives and their transmission to the SETs.

The development of a participative leadership style and increased delegation of decision making to the SETs.

The convening of an SET implementation steering committee (which will include the maintenance manager).

Foremen

Improvement of their knowledge of how multi-skilled SETs should function.

Amendment of their leadership style to complement the operation of SETs.

Development of their skills in management of change.

Involving them in advising the steering committee on the implementation and operation of SETs.

Tradesmen

Development of necessary add-on skills for multi-skilled operation.

Setting up procedures for maintenance/production shift workers to take part in joint problem-solving activities.

Promotion of shop floor understanding of the reasons for organizational change.

(k)   The maintenance foremen are the key to the successful implementation of the SETs. It is important that they are involved in this at an early stage:

- advising senior management on the duties of the SETs,
- assisting in programming the transfer of their own duties to the SETs,
- assisting in selling the concept to the SETs.

The SETs must have a clear understanding of the boundaries of their responsibilities. The foreman must take a flexible approach to his traditional duties, helping the SET when needed but otherwise standing back. During the transition stage he should act as facilitator, trainer and adviser to one or more of the SETs.

(l)   As a technician adviser or a planner, giving technical and planning support to one or more of the SETs.

Exercise 4 Food processing plant

(a)   1.   Main problem areas:
Low availability of equipment coupled with high maintenance cost.

High maintenance cost caused by:

- low utilization of tradeforce through mis-match of first line workload to first line resources,
- poor inter-trade flexibility,
- poor operator-maintainer flexibility,
- poor training,
- strict trade demarcation,
- poor work planning systems coupled with a reactive workload caused by poor preventive maintenance.

Low availability caused by:

- little or no preventive maintenance,
- low quality corrective work caused by lack of ownership at tradeforce level,
- poor documentation and history,
- polarization between production and maintenance,
- no attempt to build teams or introduce the ideas of self-empowerment.

2. The condition of the plant has been allowed to get out of control because in most areas there is no maintenance strategy other than operate-to-failure. Where there is a stated preventive strategy, e.g. in group C, it is not being carried out because of the dominance of priority corrective work. This situation has resulted in the organization evolving to cope with reactive maintenance; i.e. it has become a reactive rather than a pro-active organization. In such situations it is difficult to achieve high levels of labour efficiency. These problems are further compounded by such factors as inflexibility and poor work planning.

To improve the situation two main tasks need to be accomplished.

- Improving the condition of the plant via an injection of corrective maintenance resources. This will certainly require assistance from contractors and the equipment manufacturer.
- It will then be necessary to hold the condition at the improved level by adopting a new life plan for the units and an appropriate maintenance schedule for the plant. Resources are therefore required, at engineer and supervisor level, to design and implement the necessary life plans, procedures and systems, and at tradeforce level to carry out the necessary preventive maintenance tasks. This can be accomplished only by organizational change. It is clear that if the mis-match between the first line workload and the shift resources were to be corrected then resources would be released for a planned maintenance group. Similarly, a reorganization of

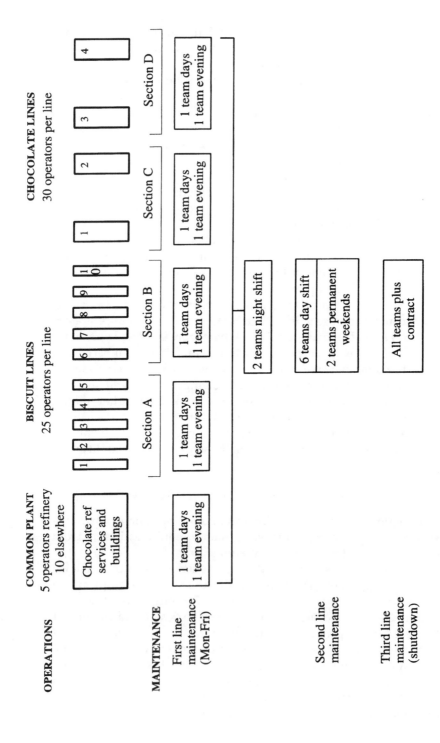

Figure 7.20   Revised resource structure for food processing plant

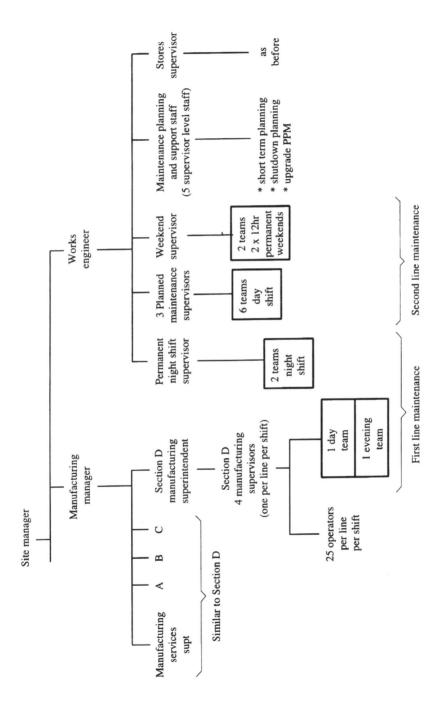

Figure 7.21   Revised administrative structure for food processing plant

the administrative structure to match international benchmark levels should release the necessary engineers and supervisors to inject the planning input.

(b) 1.   *A modified resource structure.* The information given in Figure 7.17 shows that it is important to retain some form of shift cover. It is also important, however, to match the shift resource to the workload shown. Figure 7.20 shows a modified resource structure where the shift resource has been kept de-centralized (i.e. equipment-specialized) because of the need to build production–maintenance teams and because of the specialized nature of the work. There are therefore five fitters on each main shift. The overload on days can cascade to the second line resource. A similar approach is used in the case of the electricians and wrapping fitters. The centralized night shift teams have been retained but reduced to two teams to match the workload. The reduction in the shift resource allows the creation of second line day-shift teams to handle the second line work (mainly planned maintenance). In addition, a permanent weekend group has been established, made up of two teams working 2 x 12 hour shifts (with overtime allowance, this provides a full week's work). These second line teams are centralized and work throughout the plant by job priority. The annual shutdown resource is also centralized and made up of the internal resource plus contract labour as necessary.

2.   The main move (see Figure 7.21) has been to create process-orientated manufacturing units – to which the shift maintenance teams report – in each plant section. The manufacturing superintendents and supervisors would include in their number some of the previous maintenance staff. The release of several maintenance superintendents and supervisors has provided supervisors for the second line crews and for the creation of a maintenance planning and support section reporting to the works engineer. This is the initial modification. With time, further improvements can be made as a result of introducing:

●   inter-trade flexibility,
●   operator-maintenance flexibility,
●   the ideas of self-empowerment.

8
The key maintenance system: work planning and control

Introduction

The previous six chapters have dealt with the problem of ensuring that the right level of maintenance resource is available, appropriately located and correctly directed in order to meet the workload. This can be regarded as the *statics* of maintenance management. What must now be discussed is its *dynamics*, that is the design of systems for information handling and decision making. Of these, the key component is the system for *work planning* (which, for the purposes of this book we shall take to also encompass *work scheduling and controlling*). It is this that provides much of the information for the other maintenance management activities.

The fundamentals of work planning

The work planning system defines the way in which maintenance work (preventive, corrective and modification) is planned, scheduled, allocated and controlled. Its function is to ensure that the right resources arrive at the right place, at the right time, to do the right job in the right way. Its design should aim to get the right balance between the cost of planning the resources and the savings – in direct and indirect maintenance costs – that result from employing them.

The basic steps involved in the planning and execution of any job are shown in Figure 8.1. The level of administration and systems needed to aid this process will depend on the size of the job (the manpower, spares, and time needed) and its characteristics (e.g. scheduling lead time, which could be zero for emergency work).

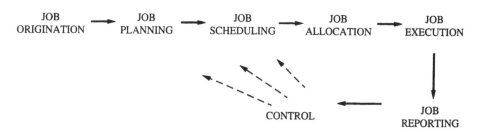

Figure 8.1   The basic steps of maintenance work planning and control

In the case of a small emergency job, for example, all of the steps indicated in Figure 8.1 may be taken by an individual tradesman. The important contribution of systems in such a situation is quickly to provide him with the essential information for planning (spares requirement, drawings, instructions) and scheduling (priority, what else is on? what else can be done at the same time?).

In the case of scheduled overhaul of a large unit, each step of Figure 8.1 may be taken by the appropriate specialist – planner, scheduler, tradesmen, clerk and so forth. A co-ordinating organization, and systems to provide it with information on jobs, planning and scheduling is then needed (see Chapter 9).

Modelling the operation of a work planning and control system

Figure 2.4 depicted a work planning system as a horizontal information and decision-making structure, the accompanying text stressing the need to design the system around the resource structure and emphasizing the planning roles at supervisor level in the administrative structure. The need to visualize the work planning system as an integral part of the overall organization was shown. This will be now be illustrated, using as an example the food processing plant (FPP) first outlined in Chapter 1.

The weekday and weekend resource structures for the FPP were shown in Figures 1.6 and 1.7, respectively. The weekday shift group carry out mainly first line work and the weekend group the scheduled second line work (see the workload diagram of Figure 1.5). The weekday administrative structure is shown in Figure 1.8. The weekend structure is headed by one of the shift supervisors (on a one-in-four rota) and aided by an electrical supervisor. Figure 8.2 shows, superimposed on the administrative structure, an outline of the work planning system. This once again serves to emphasize the need to visualize work planning as an information and decision-making system operating mainly at the supervisor level.

A more conventional way of modelling the work planning system for the FPP is displayed in Figure 8.3 (which should be looked at in conjunction with the work flow model shown in Figure 3.2). This shows the flow of work and information between the trade groups, their supervisors and the planning office and indicates that there are three distinct levels of planning:

- first line (shift)
- second line (weekend)
- third line (major shutdown) (not modelled).

The shift work planning system

The function of the shift trade group is to carry out the emergency maintenance and some minor scheduled tasks during the normal weekly shifts. This work

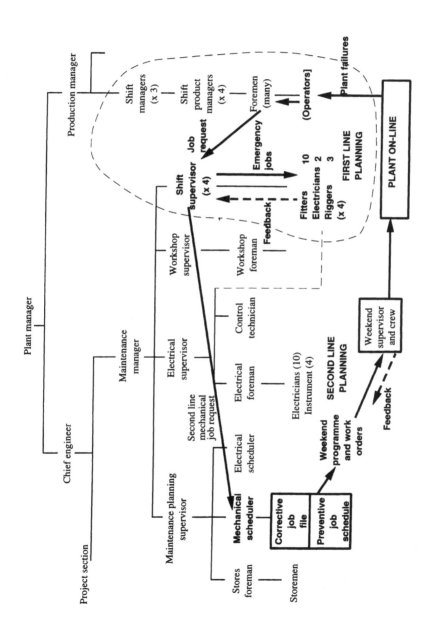

Figure 8.2   Outline of work planning system

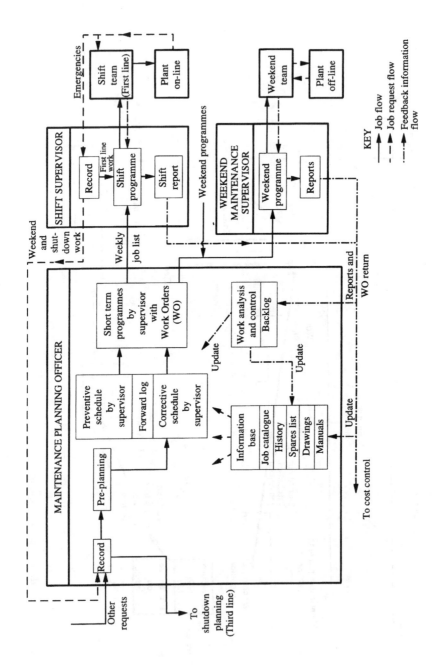

Figure 8.3   Work planning system, food processing plant

was shown as first line in Figure 1.5 and was described in detail in Chapter 3. It requires what is essentially a reactive planning system centred around the shift supervisor. The main problem is to identify priorities, especially for those jobs that should be passed back to the planning office for weekend execution. A simple way of doing this is to define all jobs that need to be carried out within twenty-four hours as first line and to prioritize them on safety and economic criteria into 1(a)s and 1(b)s. The shift crew will also deal with the minor preventive, 1(c), jobs. In general these will be sent through on a simple list from the planning office and fitted into the programme as convenient. The major shift planning requirement is to quickly obtain information on spares, drawings, safety, and job methods – hence the usefulness of a computerized maintenance information base (see Figure 8.3).

MAINTENANCE REQUEST No. D 0353				Week No		Classification	
Plant Description		Date					
Plant Location		Time	a.m. p.m.	Analysis		Plant No	
Requested By		Required By					
Defect/Work Required	IS PRODUCTION STOPPED?		YES NO	Tool BDN	1		Tick one box
				M/c Running	2		
				M/c Breakdown	3		
				Not Applicable	4		
Cause				Wear and Tear	1		Tick one box
				Accident and Misuse	2		
				Component Failure	3		
				Job Report	4		
				Not Applicable	5		
Action Taken							

Tradesman's Signature	Date	Clock No.	Time ON	Time OFF	Total Hours		Rate	£ Amount	p
					HRS	DEC			
Maintenance Foreman's Signature								HRS	DEC
							Repair Time		
For Office Use Only							Down Time		

Figure 8.4   Typical work order

In general, the main use of work orders has been to provide information on costs (see Figure 8.4) rather than to co-ordinate work and indicate job methods. Several separate work orders can be used for bigger jobs and standing work orders can be drawn up to cover the complete shift per man for small jobs.

One of the main difficulties on shifts is collecting good quality failure and repair data. At worst, it is totally lost and at best the more important events are captured on some form of individual or group shift log. Paperless computerized information systems are becomingly increasingly used in these situations. One of the planning difficulties in the case of shift groups is the co-ordination of resources, between shifts, for emergency jobs which take longer than a single shift. One obvious way of facilitating this task is to allow the shift supervisors a short overlap for updating each other. Some companies frown on this practice, however, if it increases overtime.

The weekend work planning system

The link between the planning systems for shift work and for weekend work is the planning office. Any job which, in the shift supervisor's judgement, meets the priority and planning guidelines for second line work – mainly weekend work – is referred back to that office. Modification and corrective work from other sources, e.g. from the production or engineering departments, that needs maintenance resources for its execution is similarly referred.

The usual procedure is that any incoming job has to be notified on a maintenance work request order (indicating such basic information as requester, job description, priority and so on). Such jobs can be requested by telephone with hard paper back-up, via planning meetings or by direct input into a computer system. The information on preventive jobs will already be held by the planning office system as part of the previously agreed preventive schedule.

The main function of the weekend planning system is planning and scheduling the weekend work. Part of the initial job-recording activity is to identify those jobs that are best carried out within the shutdown plan and to enter them into the job list for this (see Chapter 9), all other work being entered into the weekend system. The main function of pre-planning is to ensure that any long lead-time resources required, e.g. spares, are identified and ordered before the relevant job is entered into the schedule of outstanding tasks. This, in turn, requires an understanding of the job procedure – in outline at least. It is important, therefore, that the system identifies who is responsible for ensuring that such understanding is communicated, i.e. for specifying the job method. In the case of the FPP this was done jointly by the planner and the supervisor, some procedures being already recorded in the job catalogue (part of the plant information base). After pre-planning, the corrective jobs, with estimates of their durations and tradeforce requirements, can be entered into the job schedule, where they are filed according to supervisor, plant unit and status (on-line, off-line, etc.), priority and approximate week of execution.

It can be seen that Figure 8.3 separates the preventive from the corrective schedules. Modern software packages, however, are based on a combined job list and schedule – each job being identified, categorized and scheduled via a code. It will be appreciated that the job list is a record of all the outstanding work and that it can be presented as a histogram of future resource demand (categorized by job priority) and compared with the forecast of weekend completion of such work. This is an essential part of maintenance work control.

At each planning level there is a need for an appropriate work planning meeting. The shift maintenance system supervisors should therefore participate in the daily production meeting which, among its other activities, should review plant availability and any failures over the previous twenty-four hours. Similarly, there is a need for a weekly maintenance planning meeting (involving supervisors and manager) which liaises with the weekly production planning meeting (which should include a maintenance representative). The function of the weekly maintenance meeting is to review outstanding work and plant priorities and to decide on a weekend programme.

In the FPP the designated weekend supervisor – responsible for the allocation, quality and control of the weekend work – would be a member of the weekly maintenance planning meeting, the weekend programme having been agreed by the preceding Thursday. It would be his or her responsibility to carry out the secondary planning and job allocation, i.e. to check spares, drawings, work orders, methods, safety permits, co-ordination with other trades (electrical, production, cleaners), etc. – the necessary manpower, including contract labour, being agreed in conjunction with the planner.

Feedback of maintenance data for control

One of the responsibilities of the maintenance supervisor (shift and weekend) is the feedback of maintenance information. Such data can be captured on work orders, weekend or shift logs, or event reports – perhaps via direct computer input. The typical requirement might be for information on:

- additional work needed, or work not completed,
- resource usage (trades – with job times, spares usage, etc.),
- work carried out,
- overhauls – in case study format,
- causes of failures or potential failures, etc.

Some of this is needed for work control, some for updating the information base and some for other purposes such as control of maintenance cost, of plant reliability, of work planning effectiveness, etc. (see Chapter 10). It is for these reasons that the author describes *work planning* as the *key maintenance system*. This chapter will restrict itself to the use of such data for work planning and control purposes, i.e.

- *for the information base* (see Figure 8.3), an essential aid to work planning, the effectiveness of this database depending, however, on its continuous update – e.g. on the input of new standard jobs into the job catalogue;

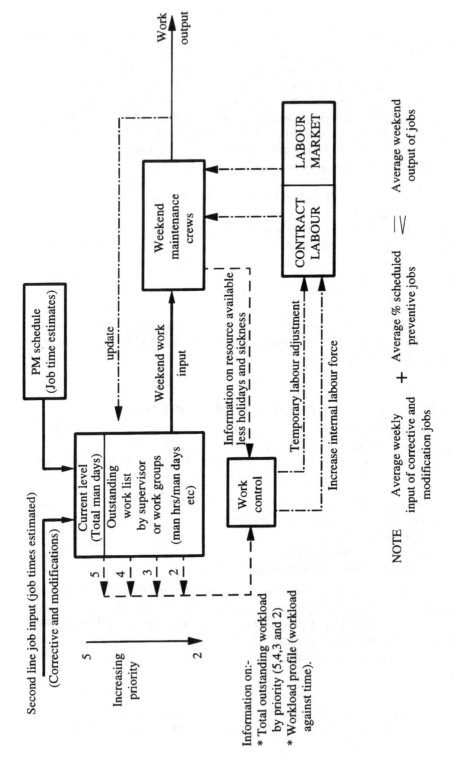

Figure 8.5　Principles of work control

- *for work control*, achieved via the feedback of information on jobs completed, time taken, jobs deferred, etc. (this allows the main job lists/schedules to be updated, a view obtained of the outstanding work – forward log and backlog – compared to the completion rate (see Figure 8.5) and extra resources (contract or in-house) to be brought in if needed).

Work control should also include the de-briefing procedure necessary after any unit overhaul, or even after the weekend programme, the purpose being to ask 'Did we do it right?', 'Could we do it better?'

Comments on work planning and work control

The FPP was chosen to discuss maintenance work planning because it has several features that render the operation of its work planning system straightforward:

- The first line work is done by shift teams, and the second line work by weekend teams, so the chance of the first line work interfering with the execution of the scheduled work is minimized. Where a single group carries out both categories of work, the planning becomes considerably more difficult. Questions arise concerning how much planned work should be committed into the weekly programme. This is sometimes estimated as follows:

$$\begin{bmatrix} Scheduled\ workload \\ for\ the\ next\ period \end{bmatrix} = \begin{bmatrix} Workload\ capacity\ of \\ trade\ group\ for\ the \\ next\ period \end{bmatrix} - \begin{bmatrix} Average\ level\ of \\ non\text{-}plannable\ work \\ likely\ in\ the\ next\ period \end{bmatrix}$$

This might then allow the programme to have both a *committed* and a *flexible* work element, in which case priority rules must be clearly stated.

- The role of the planners can be clearly defined, and their use justified, as regards the weekend work. This is because the shift supervisors cannot carry out the planning role as part of their main responsibilities. They act as weekend supervisors on only one week in four – they and the FPP need the ongoing planning back-up. Situations do occur, however, where the supervisor acts as 'planner', the leading hand as 'supervisor' and the planner as 'clerk'. In other words, there may be second line situations, especially where good computerized maintenance systems are in use, where a specialized planner may not be required.

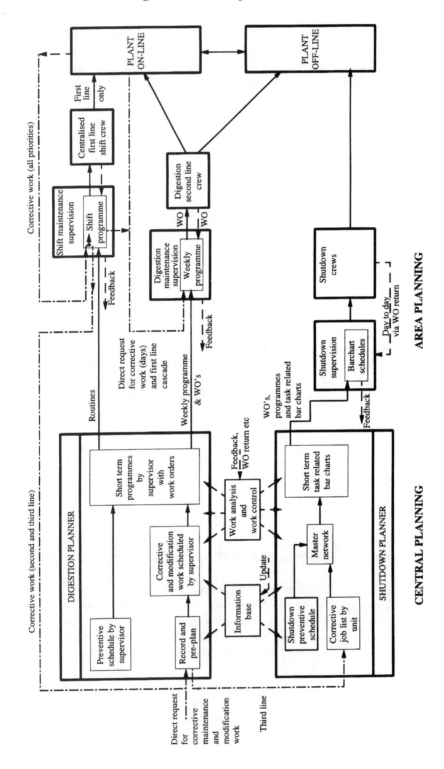

Figure 8.6   Work planning system for an alumina refinery

Guidelines for the design of work planning systems

Consider the work planning system shown in Figure 8.6, which has been designed around the resource structure of Figure 8.7 (a simplified version of the structure recommended for the alumina refinery, see Figure 4.14).

Figure 8.7 shows three resource levels corresponding to the centralized shift crews (first level), plant-located de-centralized day crews (second level) and centralized shutdown and plant service crews. This example draws attention to several important aspects regarding the design of work planning systems:

(i) The work planning system should be designed around the resource structure taking into consideration the levels of resource and the characteristics of the workload. The flow of work can be mapped as in Figure 3.2. If, as in the present example, there are three levels of resource then it is necessary to have three corresponding levels of work planning.

The scheduling lead time of the emergency work dealt with by the shift crews is short and planning centres around the supervisor and tradesmen. The main requirement from the documentation system is to provide the necessary information quickly.

With the second line crews scheduling lead times vary from twenty-four hours to several weeks and the jobs are tackled by priority. Planning should centre around the respective planners – e.g. the planner of Figure 8.6, who provides a weekly list of committed work taking into account the first level spillover. His main function is to assist in the planning of individual jobs and to keep the schedules and work lists up to date. In addition he assists in the co-ordination of multi-trade jobs and provides general clerical back-up. In the Figure 8.6 example planning and scheduling is a joint planner–supervisor effort. It must be emphasized that with the most up to date computerized planning system the area planning is often carried out by the supervisor/adviser especially if the areas have become self-empowered.

Shutdown planning for major parts of the refinery exploits the presence of redundancies, which enable plant (a boiler, kiln, etc.) or a process channel (a digester stream, for example) to be taken off-line – perhaps reducing capacity but never shutting down the complete refinery. The centralized shutdown crews and the service crews then go to the area concerned and supplement the local resource, the planning centre of gravity being close to the shutdown planning office.

The function of the shutdown planner is to ensure that the shutdowns are scheduled so as to smooth the workload of the centralized resource. This requires close co-operation with the

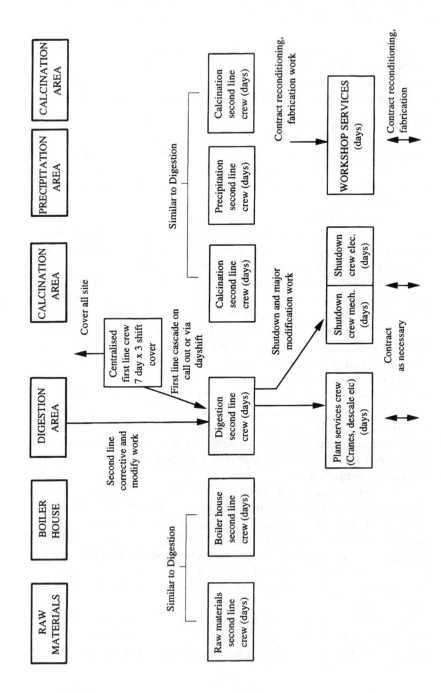

Figure 8.7   Simplified version of proposed resource structure for an alumina refinery

production planners. In addition, the shutdown planner ensures that the jobs are properly planned (in particular, long lead-time spares are identified) and the resources properly co-ordinated. The planning lead time for shutdown work is often in excess of three months (see Chapter 9).

(ii) The design of the overall planning system should allow job requests to flow to the *best* level for planning, scheduling and controlling them, e.g. shift supervisor level for emergency work. In the example, jobs specific to the digestion stream that *do not* require the plant to be off-line should be handled at digestion planning level, those that *do*, at shutdown level. By 'best' is meant where it is easiest to decide on priorities, plant status and job methods.

(iii) A set of job priority rules – which should indicate the initial planning level (e.g. emergency maintenance to be planned at first line supervisor level) – should be defined and clearly understood by maintenance planning and supervision and by production.

(iv) The function, responsibility, and lines of communication between each of the planning levels must be clearly defined.

(v) The right information about the workload and the resources must be available to the designated planner of the job at the right time.

(vi) The planner must have the authority (or access to it) to take the decisions (e.g. allocation of priorities) which affect the workload and resources.

Safety aspects

There are many aspects of the maintenance work planning system that are important to the safety of personnel and of plant operation.

- Plant isolation procedures, including safety permits and the isolation tagging system, are essential in a hazardous area and must be incorporated into the work planning system.
- Standard job procedures, including methods, spares used, tools, and safety procedures, should be mandatory in hazardous plants.
- All major overhauls should be preceded by a discussion of safety procedures and requirements.
- All contract labour coming on to site should be subject to job quality checks and should go through a safety induction programme appropriate to the site and to each of its areas.
- All spare parts, whether new or reconditioned (internally or externally), should be subjected to a quality assurance programme, particularly if they are destined for hazardous areas or functions.
- Formal procedures must be established for updating drawings and manuals after plant modification – and for communicating

manufacturers' requirements or modifications to the holders of such information.

An exercise in work planning

This exercise is based on the system that was modelled in Figure 8.6 and then briefly discussed – a system for planning and controlling the maintenance work of a large process plant. The resource structure, which was also outlined (see Figure 8.7), comprises:

- centralized shift crews undertaking the first line work (emergency and minor corrective work, preventive routines);
- several de-centralized plant-based area crews, on days, dealing with the second line work (all the plant-based jobs in their own areas, other than the major shutdowns);
- a shutdown trade group made up of internal and contract labour and sized to the shutdown workload.

Each of the crews has its own supervisor. Various de-centralized plant-based planning offices (e.g. for the Digestion plant) have been set up to assist the area maintenance supervisors. There is also a centrally located shutdown planning office for the management of major overhauls.

The planning system of Figure 8.6 is fully computerized, having a comprehensive maintenance information base and an information analysis and control module. In addition this system is integrated with stores and invoicing control and it interfaces with the company's costing system. Although this was not shown, the information captured on the work orders (e.g. man-hours expended) and on the stores requisition (e.g. materials used) is an essential component of the costing system.

Task 1

Explain why each of the following statements regarding Figure 8.6 is probably correct.

- For the shift crews the centre of gravity of maintenance work planning lies well towards the shift supervisor.
- Both for the shutdown crew and the workshop-based crews the centre of gravity of work planning lies well towards the shutdown planning office.

Task 2

Some of the planning and scheduling for the second line de-centralized crews is carried out by the area supervisors and some by the area planning supervisors.

(a)   Explain why it is important that all jobs are entered into the system (before their execution where possible, but perhaps afterwards in the case of emergency work).

(b)   Identify the work that, on the one hand, is best planned and scheduled by the area supervisors and, on the other, by the area planning supervisors.

(c)   Should jobs that come via the area planning supervisors be planned and scheduled completely? Should this go so far as the allocation of jobs to specific tradesmen?

(d)   Explain the importance, for the second line area supervisors, of a job priority system.

(e)   Why is it important to estimate job times before adding the jobs to the computerized job lists and schedules?

(f)   Describe what is required from the work planning system if it is to ensure that the plant will operate safely and that maintenance will be carried out safely.

(g)   Management are aware that the success of the maintenance control systems will depend in part on successful collection of data from the work planning systems. List some of the difficulties that may be encountered when trying to achieve this.

Guideline answers to the exercise

Task 1

First level work planning deals with jobs that are of high priority (planning lead time less than twenty-four hours) or are small ones that are not subject to planning input from outside (financial sanctions, the ordering of spares, etc.). The shift supervisor therefore plans and schedules such work.

In the case of shutdown work (and also of work undertaken by the workshop-based crews) the planning lead time may be as long as a year, so a great deal of financial and work planning is essential. Many of the planning and resourcing decisions can be taken only with information which is not available to area supervisors.

Task 2

(a)   In order to ensure the capture of data – on man-hours needed and expended, on cost of materials and on plant history.

(b)   *Area supervisors* – jobs that do not require information or resources from outside, or are not subject to external financial sanctions. As long as such jobs are entered into the system they are adequately visible to the area planning supervisor.

Area planning supervisors – jobs that call for co-ordination with other departments, (including Operation or Production), that involve the use of a common, centralized, resource, or are subject to financial sanction.

(c)   No. The area planning supervisors should always leave detailed and shorter-term planning and allocation to the area supervisors, who are much more aware of the capability of their resources and of the short-term planning characteristics.

(d)   It ranks the jobs according to their importance to production and/or safety, which facilitates estimation of their scheduling horizons.

(e)   Because it enables the workload to be profiled and compared with the available resources, an essential part of work control which also promotes better short-term planning.

(f)
- Plant isolation procedures
- Standard procedures for hazardous jobs
- Safety review meetings before major overhauls
- Safety inductions for new, and for contract, workers
- Quality assurance review of all new and reconditioned parts
- Updates of drawings and manuals if plant is modified
- Plant history recording, and analysis to anticipate potential safety problems

(g)
- Tradesmen's resistance to providing data (habitual reluctance to co-operate, lack of time, dislike of writing, etc.)
- Poor quality of data (due to insufficient understanding of data needs, poor definition of requirement, etc.)
- Trade union insistence that data collection must be financially rewarded
- Lack of training in the use of the documentation system
- Poorly designed documentation systems which do not facilitate effective use of data and which therefore inhibit any enthusiasm for its collection
- Lack of commitment – on the part of supervisors and other managerial staff – to data collection.

9

Management of plant turnarounds[†]

Introduction

Major plant turnarounds[*] pose problems of work planning and scheduling that are quite different from those presented by the ongoing work. The principal organizational characteristics of the typical third line workload that they generate are as follows:

- a large peak in resource requirement – which has to be met by an influx of labour from elsewhere;
- a multiplicity of inter-related activities, all of which have to be co-ordinated if the work is to be completed on time and to cost;
- a long lead time (often many months) for scheduling the work;
- large cost penalties should the planned duration of the shutdown[**] itself be exceeded.

Planning and controlling the work for a major shutdown is therefore an exercise which is quite separate from the corresponding task for the ongoing work (although it relates to it). This can be seen from Figure 9.1, which models the complete work planning system for the alumina refinery. All jobs originating in the planning system for ongoing work (or elsewhere) that can only be undertaken during the major shutdown are transferred into the shutdown corrective job list. The sum total of these (which can be many thousands for a large overhaul[***]) requires many months of planning and scheduling and is expressed as a shutdown programme on a master network (based on Critical Path Analysis – see Appendix 1). This network can then be broken down into a series of short-time-frame bar-charts with accompanying task specifications. It can be seen from the figure that a number of functions are common to both the ongoing and the shutdown work planning systems, namely information base, work analysis, work order system, cost control.

[†]Contributed by T. Lenahan, with additions by A. Kelly

[*] *Turnaround*: an engineering event that takes place during a plant shutdown and involves the inspection, overhaul and, where necessary, the modification of existing equipment and the installation of new equipment.
[**] *Shutdown*: the period of time from the moment the plant is taken off line to the moment it is brought back on line.
[***] *Overhaul*: a comprehensive examination of a plant, or a major part of it, and its restoration to an acceptable (or desired) condition.

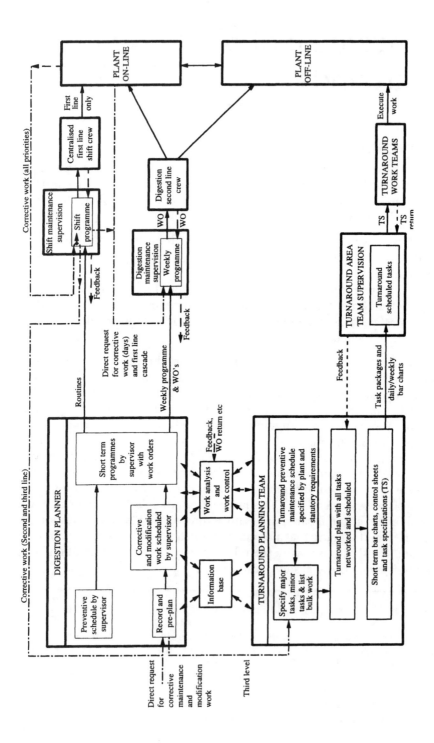

Figure 9.1 A work planning system for an alumina refinery

In a chapter of this length the most that can be accomplished is an outlining of the general procedure for managing turnarounds and overhauls – and hence the complete shutdown – and a highlighting of some of the main problem areas. The subject is a large and complex one, however, and worthy of a textbook in its own right.

Turnaround methodology

A general methodology for managing turnarounds is outlined in Figure 9.2. The key strategic decisions and their timing relative to the four principal phases (each of which is divided into sub-steps) are indicated along the top of the figure.

Many variations of this methodology have been observed. Figure 9.3, for example, is a turnaround schedule for a small agrochemical plant. In this case the planning lead time (i.e. the total planning time required up to the start of execution of the work) is about two months. A turnaround for a 350 MW boiler–turbo-generator might well demand a planning lead time of a year. The magnitude of the planning lead time and effort depend on the size and complexity of the turnaround and of the previous experience of similar or identical exercises. Major plant overhauls are usually much the same in one year as in another and this therefore makes the initiation and preparation phases that much more straightforward; this is the principal difference between turnarounds (where most of the work is maintenance) and major plant projects which involve the building and installation of new plant.

Phase 1: Initiating the turnaround

Forming a policy team and appointing a turnaround manager

It is the responsibility of the senior management of the company to form a policy team made up of the managers who have a stake in the turnaround and the ability to make decisions. This team must ensure that constraints, objectives and policy for the turnaround are well defined. Arguably, the most important decision made by the team will be the appointment of a turnaround manager to act as their agent because, once appointed, this individual will control all phases of the task, while the team will assume the tripartite role of facilitation, monitoring and endorsement.

The turnaround manager should chair a regular meeting of the team at which progress will be reported and constraints, objectives and policy discussed on matters including timing, duration, workscope (including major tasks), materials, resources, contractor issues, pre- and post-shutdown work, costs, safety and quality.

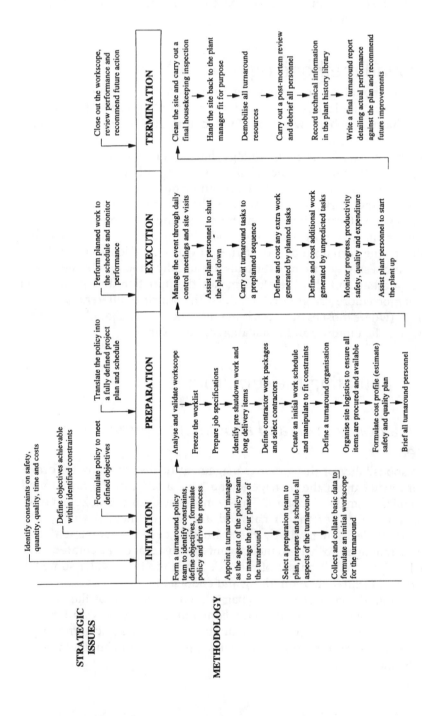

Figure 9.2   Turnaround methodology

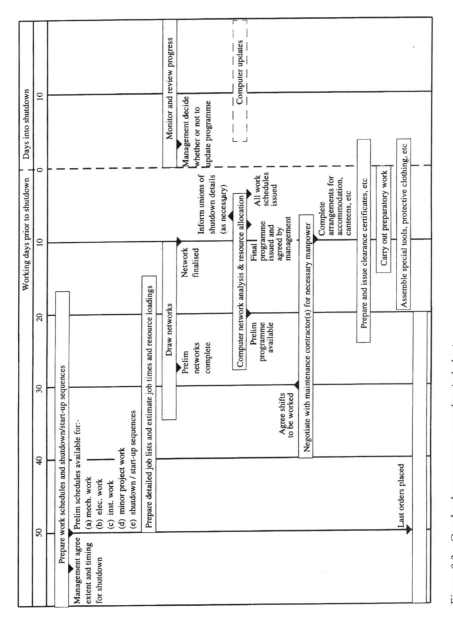

Figure 9.3   Overhaul programme, agrochemical plant

Key policy decisions for the initiation phase

Two of the most important decisions that have to be made during the initiation phase involve the formulation of the turnaround objective and the timing of the plant shutdown. The objective needs to be compatible with the overall maintenance objective. For example in the case of a base load boiler–turbo-generator unit (see Figure 9.4) the maintenance objective might be:

> to minimize the sum of (i) the income lost due to planned outage, (ii) the income lost due to unplanned outage, and (iii) the direct cost of maintenance.

The turnaround objective might then be:

> to complete the agreed maintenance work within the agreed shutdown duration at minimum resource cost, and while meeting all relevant safety standards.

This general statement could also be supplemented by the setting of more detailed targets and standards for workload, work quality, plant condition, safety duration and cost. This objective is feasible because the generator would be shut down during the annual period of low demand and there would be little to gain by reducing the shutdown duration. For production-limited chemical plant on the other hand the objective might be very different, putting the emphasis on the reduction of turnaround duration – at the expense of high maintenance cost. Clearly, the turnaround objective must be established at an early stage.

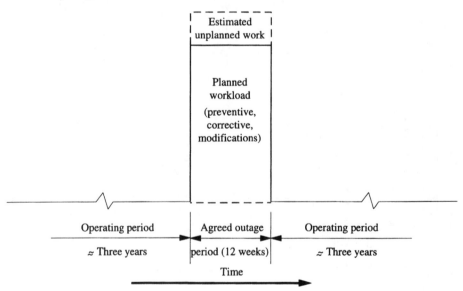

Figure 9.4   Operating pattern of a base load generating unit

The timing of major plant shutdowns (e.g. whether they should occur at some fixed interval or should be usage or condition based) and their preventive work content are determined to a large extent by the maintenance strategy (see Book I). Base load boiler–turbo-generator plant, for example, typically come off every three years or so as part of a station schedule with a twenty year planning horizon. The exact timing of the shutdown will be a function of the annual fluctuation in electricity demand, and other factors such as the need for grid maintenance. Although the frequency of the shutdown is time based (inevitably so in this extensive planning context) the preventive work content is determined in part via inspections.

Selecting a preparation team

The turnaround manager is responsible for selecting a preparation team comprising:

- *a preparation engineer* – managing the team on a day-to-day basis, validating the workscope and delegating other work to the planners;
- *a planning officer* – supervising the planning team, assisting the preparation engineer to validate the workscope and creating the initial turnaround plan;
- *a planning group* – gathering basic data on the plant (e.g. drawings, specifications, plant standards), specifying tasks and providing all supporting documentation such as specifications of welding and pressure test procedures;
- *a site logistics officer* – organizing storage, supply and distribution of materials, equipment, cranage, transportation, services, utilities, accommodation and facilities.
- *a site logistics team* – of storemen, drivers and other semi-skilled personnel who work under the supervision of the logistics officer.

Collecting the job lists and other data

In order to prepare the turnaround plan the preparation team will need to gather a substantial amount of data from a number of sources. To this end, the manager should set up a series of meetings (see Figure 9.5) which will run through the preparation phase until all relevant data has been obtained.

Phase 2: Preparing the turnaround

During the preparation phase the worklists and basic data obtained from the plant team should be transformed into a turnaround plan which covers all requirements, including task specifications, work programme, duration, cost, manpower, safety and quality (see Figure 9.6).

PLANT STANDARDS

Purpose - to ensure the turnaround is executed within plant standards

Agenda
-Plant SOPs
-Engineering standards
-Safety standards
-Quality standards
-Any other local or special standards

WORKLIST

Purpose - to ensure the workscope is properly defined

Agenda
-Define requirements
-Challenge work
-Eliminate unnecessary and duplicated work
-Check material specs
-Finalise worklists
-Freeze worklists

MAJOR TASK REVIEW

Purpose - to ensure all requirements for major tasks

Agenda
-Information gathering
-Specify task
-Write task procedure
-Specify requirements
-Task hazard study
-Rescue plan
-Preparation plan

PROJECT REVIEW

Purpose - to ensure all projects are properly defined and resourced

Agenda
-Specify project
-Define key dates
-Finalise drawings and documentation
-Materials/equipment
-Specify interactions with turnaround work

INSPECTION REVIEW

Purpose - to ensure all inspection requirements are resourced

Agenda
-Specify inspection list
-Consider deferments
-Inspection techniques
-Types of inspectors
-Safety and access requirements

CONTRACTOR REVIEW

Purpose - to ensure all contracted out work is properly managed

Agenda
-Work packaging
-Types of contract
-Contractor availability
-Invitations to bid
-Contract award
-Contractor control and management on site

SPARES REVIEW

Purpose - to define items to be supplied within plant standards

Agenda
-Spares availability
-Check specifications
-Withdrawal from plant stores
-Issue and control
-Return of surplus plant stores

SITE LOGISTICS

Purpose - to ensure the turnaround site is properly organised

Agenda
-Site plot plan
-Materials/equipment
-Stores/layout areas
-Craneage/transport
-Services/utilities
-Accomodation
-Facilities/personnel

SHUTDOWN/STARTUP

Purpose - to integrate SDSU plans into the turnaround plan

Agenda
-Define shutdown plan
-Define startup plan
-Specify requirements
-Specify special gear
-Link shutdown and startup to turnaround schedule

SAFETY REVIEW

Purpose - to develop a safe system of work for the turnaround

Agenda
-Safe working routine
-COSHH
-The safety team
-Safety inspections
-Accident investigation
-Safety initiatives

QUALITY REVIEW

Purpose - to ensure all quality standards are understood and met

Agenda
-Quality system to be used
-Material quarantine
-Non-conformances
-Instrument calibration
-Hand-back acceptance

Figure 9.5   Meetings organized during preparation phase

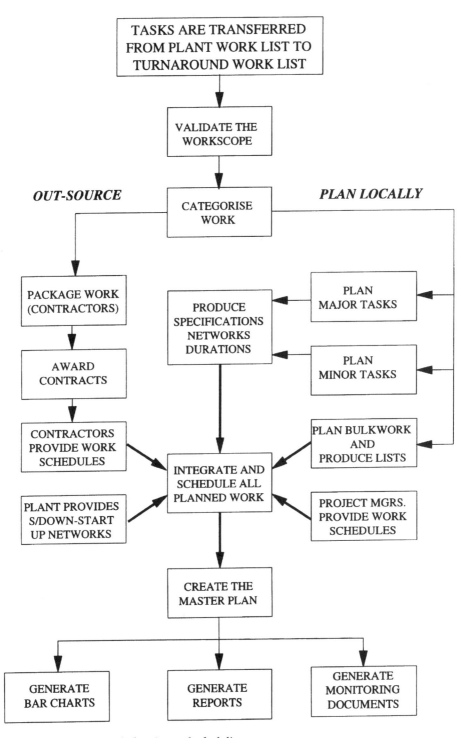

Figure 9.6 Turnaround planning and scheduling process

Analysing and validating the workscope

The worklists supplied by the plant team will be raw data containing elements of unnecessary and duplicated work (which must be eliminated), incorrectly specified work (which must be corrected) and errors in material specification, etc. During a series of meetings (see Figure 9.5) the preparation team should set out to validate the workscope and define all requirements by analysing every major, minor and bulkwork request on the worklist to ensure that it needs to be done, it is correctly specified, and that all requirements are specified.

Freezing the worklist

The validated workscope will be the foundation upon which every other aspect of the turnaround rests. Therefore, at a pre-determined date (normally between two and six months before the event) the worklist should be frozen and no further work accepted. The frozen worklist can then be transformed into the workscope that will be used to accurately calculate key indicators such as cost, duration and resourcing of the event. In reality, work will usually be requested up to and even beyond the start date of the turnaround. If the worklist is not frozen it would be impossible to accurately calculate these key indicators. Any work requested after the freeze date should be handled by a 'Late Work' routine which should be costed and resourced separately.

Preparing task specifications

The raw worklist will consist basically of three categories of work, each of which will need to be treated differently as regards planning and specification.

- *Major tasks* (e.g. overhauling a switchboard)
  Large packages of work that are characterized by long duration, high technical content, unfamiliarity or high risk. The preparation engineer should be responsible for planning and specifying both the technical and safety content of these and for producing a critical path network (see Figure 9.7).
- *Minor tasks* (e.g. washing and inspecting a small heat exchanger)
  Small packages of work of medium duration which still require individual planning and specification. These tasks are planned and specified by a planner, on a task sheet (see Figure 9.8).
- *Bulkwork* (e.g. replacement of many similar valves)
  Numerous small jobs with identical or similar requirements that can be packaged by type and listed on a bulkwork route card (see Figure 9.9).

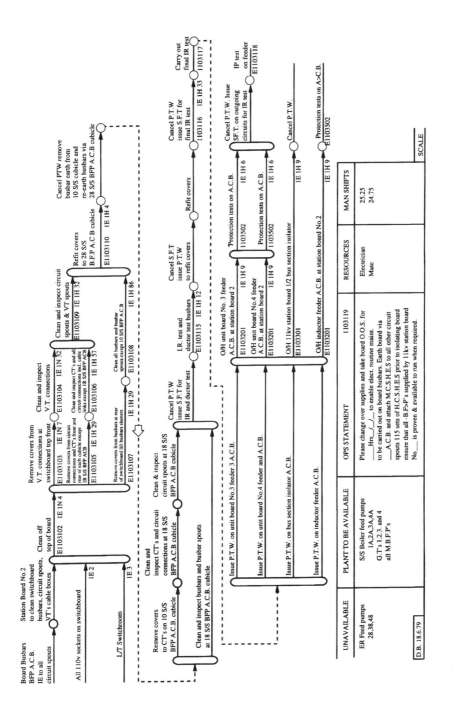

Figure 9.7 Critical path network for an 11 kV switchboard overhaul

EQUIPT NO : C-401	SERVICE	: NHTU/CRU SOUR WATER SEPARATOR	JOB TITLE	OPEN UP FOR INSPECTION
SCAFFOLDING :	*No*	CRANE : *No*	INSULATION *No*	

Sketch of equipment

B1 · B2 · B3 · B4

AN	HAVE YOU IDENTIFIED ALL SAFETY REQUIREMENTS	RESOURCES						DRN
		E	S	FT	OP	IP	LB	HRS
AO1	ERECT EXTERNAL SCAFFOLD	4						
AO2	INSTALL ISOLATION			4				3
AO3	OPEN MANHOLE			3				3
AO4	INSTALL VENTILATION FAN	2						2
AO5	02 TEST BY OPERATION & OBTAIN SAFE ENTRY PERMIT				1			1
AO6	INSTALL LIGHTING	2						1
AO7	INSPECT					1		1
AO8	INTERNAL CLEANING						3	4
AO9	FINAL INSPECT BY INSPECTION					1		1
A10	CLOSE MANHOLE			3				3
A11	DE-ISOLATE			4				4
A12	REMOVE EXTERNAL SCAFFOLD		4					4

SAFETY REQUIREMENTS (YES)	ENTRY	ELEC.ISOL.	GUARDIAN	B.A. SET
	YES	NO	YES	NO
	LIGHTS	EDUCTOR	OTHERS	
	IES - 2EAS	YES - 1EA		

LOCATION
Layout plan: CRU/B9

C110 · C110 · C401 · EI 105 (L) · EI 117 · EI 113B · EI 113A (U) · C111 · EI 106 (L) · EI 109 · C400

UPDATED ON | PLANT

Figure 9.8   Sample task sheet

BULKWORK ROUTE CARD No ....

YEAR .... PLANT.................................... UNIT/AREA .... ITEMS ....

ITEM NUMBER	ASSOCIATED PLANT NUMBER	INST/ELEC DISCONNECTION (IF REQUIRED)	REMOVED	DECONTAM	SENT TO W/SHOP FOR OVERHAUL	RETURNED FROM WORKSHOP	OVERHAULED ITEM FITTED	SPARE ITEM FITTED	INST/ELEC RECONNECTED (IF REQUIRED)	TESTED AND COMMISSIONED

Figure 9.9   Bulkwork route card

Identifying pre-shutdown work

During the analysis of the worklist some jobs, spares or materials will be identified which, because of their special nature, need to be dealt with as early as possible, namely:

- long delivery spares and materials, which must be ordered at the earliest possible date if they are to be available when required (e.g. delivery of a replacement main-compressor rotor can take up to eighteen months);
- spares which have to be prefabricated and tested before they can be installed in the plant;
- specialist sub-contractors and equipment – the more specialized, the longer their waiting list, so their availability should be checked and orders placed on them early;
- general services and utilities – such as temporary electrical and telephone cables, water and gas pipelines, etc. – must be laid well in advance of the turnaround start date.

Defining contractor work packages and selecting contractors

The decision must be taken as to what work will be contracted out, how it will be packaged and what mix of contractors will be used.

Creating the work schedule

Once tasks have been specified (and, where necessary, networked) to the appropriate standard – and all necessary materials, equipment, resources and services identified – they must be assembled into a schedule which will meet the current constraints on workload, money, time and resources. Figure 9.10 shows the elements of the schedule and the steps that need to be taken to prepare and finalize it. The logic register referred to is simply the working patterns which are initially set up in the schedule, e.g.

- bulkwork to be executed on an eight-hour shift, five days per week, only;
- minor tasks and some major tasks on twelve-hour shifts, seven days per week;
- nominated major tasks (including the critical path task) to be accomplished on a continuous twenty-four-hour cycle).

The start date and time of any given task will be dependent upon when the job is released by the plant shutdown programme and the resources available at that time. All tasks that do not lie on the critical path will have a certain amount of float time and this will be translated into earliest and latest start and finish times for each significant element of the task. This allows a measure of flexibility in the plan.

If a turnaround is large and complex it is customary to produce networks at two levels. For example, Figure 9.11 is a 'key date network' for a boiler–turbo-generator overhaul. At the level of this network an arrow represents a major section of the overhaul. Such major sections of work can be themselves represented by a detailed computer-generated network (see Figure 9.7); at this level the arrow represents a single task. The master network can be regarded as a route map of the complete turnaround, showing how all the activities are logically related. As already indicated, the networks are best converted to bar charts for the execution phase (see for example Figure A9 of Appendix 2).

Methods of creating a schedule

There are basically four methods of creating a schedule:

(i)   using the planning scheduler in the client company's computer;
(ii) using a PC and stand-alone package (such as *Microsoft Scheduler* or *PrimaVera*);
(iii) manually; using a shuffle board which allow tasks to be shuffled back and forth along a time line;
(iv) manually, using a standard programme planning sheet.

Two points should be noted, however.

(i)   Most commercial scheduling packages are not ideal for

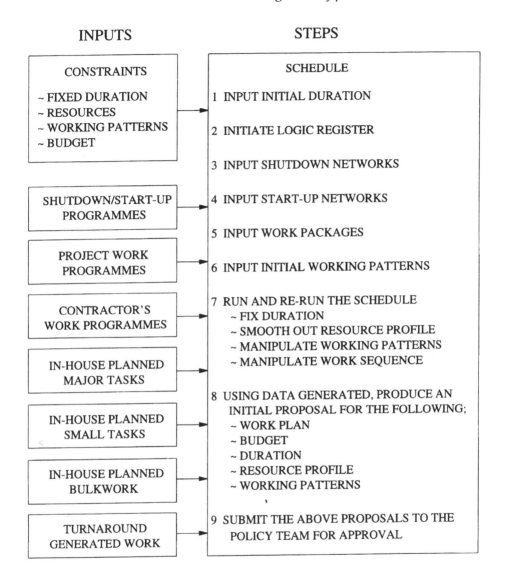

INPUTS   STEPS

| CONSTRAINTS | SCHEDULE |

CONSTRAINTS

~ FIXED DURATION
~ RESOURCES
~ WORKING PATTERNS
~ BUDGET

SHUTDOWN/START-UP
PROGRAMMES

PROJECT WORK
PROGRAMMES

CONTRACTOR'S
WORK PROGRAMMES

IN-HOUSE PLANNED
MAJOR TASKS

IN-HOUSE PLANNED
SMALL TASKS

IN-HOUSE PLANNED
BULKWORK

TURNAROUND
GENERATED WORK

SCHEDULE

1 INPUT INITIAL DURATION

2 INITIATE LOGIC REGISTER

3 INPUT SHUTDOWN NETWORKS

4 INPUT START-UP NETWORKS

5 INPUT WORK PACKAGES

6 INPUT INITIAL WORKING PATTERNS

7 RUN AND RE-RUN THE SCHEDULE
   ~ FIX DURATION
   ~ SMOOTH OUT RESOURCE PROFILE
   ~ MANIPULATE WORKING PATTERNS
   ~ MANIPULATE WORK SEQUENCE

8 USING DATA GENERATED, PRODUCE AN
   INITIAL PROPOSAL FOR THE FOLLOWING;
   ~ WORK PLAN
   ~ BUDGET
   ~ DURATION
   ~ RESOURCE PROFILE
   ~ WORKING PATTERNS

9 SUBMIT THE ABOVE PROPOSALS TO THE
   POLICY TEAM FOR APPROVAL

Figure 9.10   Creating the schedule

turnarounds because they have been designed for project
management (which tend to have long duration tasks). In addition,
they do not handle bulkwork in a satisfactory manner and
often distribute it throughout the programme to suit resource levels
(so that, for example, two valves at the same location may
be scheduled to be removed on different days when common
sense would dictate that they be removed at the same time). This

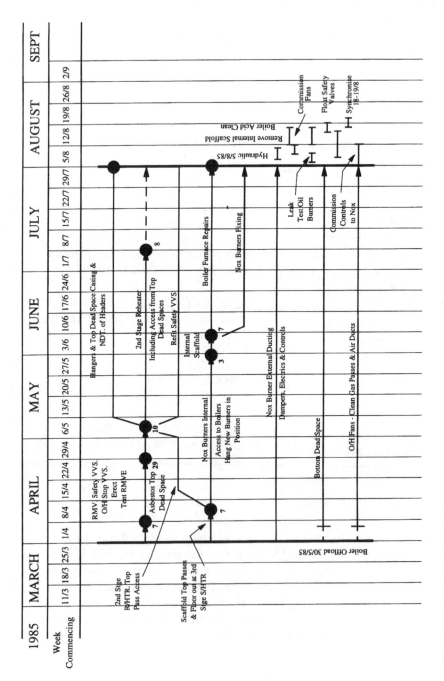

Figure 9.11   Key date bar chart

case is not as trivial as it may at first seem because, on a major event there will be many hundreds, if not thousands, of such bulkwork items. This leads to a situation where a computer may be used for scheduling tasks in general but bulkwork tasks will be scheduled manually.

(ii) Manual planning (on a shuffle board or planning sheet) is only practical for small turnarounds because the amount of detail required for a large one would swamp a manual planning format. Also, if there is a significant change of intent in the turnaround logic, the plan is difficult to change.

Optimizing the schedule

An optimum schedule balances out the constraints of workload, duration, cost and resources. The first run of the schedule will be based on raw data. For instance, it may be that the duration required by business needs is not realistic for the required workload and the available resources and if the resources were to be increased the cost would over-run the budget. A first-run schedule often exhibits an erratic manpower profile, i.e. a requirement for significantly different levels of resource on consecutive days (e.g. mechanical fitters required might number 189 on Day 1, 66 on Day 2, 12 on Day 3, 82 on Day 4, 29 on Day 5, and so on). This would have to be evened out by re-organizing the work schedule.

The turnaround manager is responsible for optimizing the schedule and then presenting it to the policy team for discussion. It may well be that this process would have to be reiterated several times before the policy team's final approval is obtained and bar charts produced showing the sequence of tasks on a daily and weekly basis. These charts will need to be updated daily during the execution phase.

Forming the turnaround organization

The form of the organization will be dictated by current constraints and policy team decisions. For example, Figure 9.12 outlines an organization, managed by a consultant turnaround manager, which dealt with three areas: one which involved a large project handled by the company's project department, another which was handled by a contractor and yet another which was handled by company personnel (both of the latter being covered by a single co-ordinator). The control electrical work was treated as a separate 'area'. The plant team handled the shutdown and start-up phases and the company supplied a team to control work quality (and who signed-off tasks when they were satisfactorily completed).

A typical organization would comprise: plant personnel with local knowledge; turnaround personnel with planning, co-ordination and

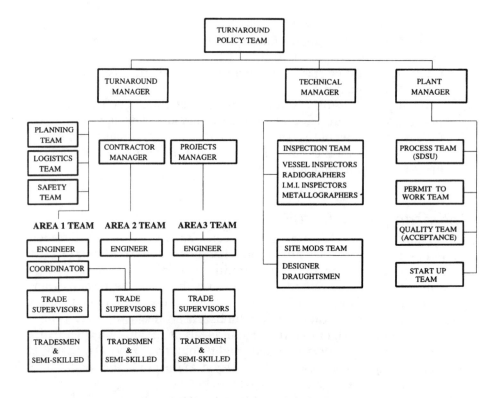

Figure 9.12   An example of a turnaround organization

management skills; technical personnel with engineering, design and project management skills; contractors with the skills and knowledge to carry out the work. Throughout the organization, control would be maintained by the practice of single point responsibility, which requires that, at any given stage of every task there will be one person who has been nominated as responsible for accepting the task from the previous stage, ensuring that the current stage of the task will be completed to the required standard and handing the task on to the next stage.

An organization that has been employed for carrying out turnarounds of certain power plants is shown in Figure 9.13. Plant-specialized core teams were set up for each major plant area (e.g. for the boilers). Supplementation of these, as necessary, from a centralized trade pool created an organization which was essentially an area–trade matrix.

Defining the site logistics

Turnaround logistics are concerned with the procurement, location and movement of all items and services required. The site logistics are

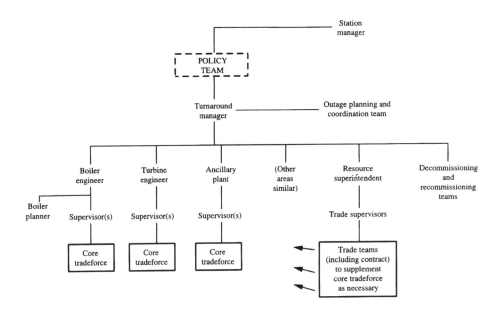

Figure 9.13   A typical power-plant turnaround organization

concerned with identifying the turnaround resources and services and their current location and organizing their movement around the site. A team under the leadership of a logistics officer (see Figure 9.12, for example) should be responsible for providing all turnaround requirements.

One method of displaying much of the information required for logistical decision-making is to draw up a plan of the site that shows the plant and its available surrounding land, and to overlay this with the locations of such things as stores, lay-down areas, water-washing bays, turnaround offices, accommodation, etc. Vital safety information can be recorded in the form of locations of non-loadbearing surfaces, prohibited or hazardous areas, fire assembly points, toxic refuges, emergency showers and eye baths. Such a plan ensures that everyone involved has a general understanding of the logistical arrangements.

Formulating a cost estimate

As soon as the main features of the workscope are known, an estimate of the cost – to within plus or minus 20% – could be generated by calculating the approximate number of man-hours required to carry out each major job, each minor job and all of the bulkwork, and then multiplying this figure by an average hourly rate. If this manpower cost were then taken to represent approximately 30–35% of the total cost, then trebling it would give the total cost of the turnaround, to within the tolerance stated. (Obviously, if any *accurate* cost information were to be available at this time it should be used in the calculations.) This figure (the *ball-park* estimate) should be calculated early on in order to give the policy team some confidence that the estimated costs are of the correct order, or to give them the opportunity to make changes if they are wildly different from the sum which has been set aside for the turnaround in the strategic budget. As harder information becomes available the estimate can be re-calculated, its accuracy increasing until a stage is reached when all significant costs are either known or can be assessed very accurately (resulting in the *refined* estimate). This stage can be reached after the turnaround plan has been approved.

Formulating a safety plan

During a turnaround the normal routine of a plant is breached, many more people than usual will be concentrated in its limited area, many will be strangers to the plant and its hazards, and most will be working under pressures of time – all of which has the potential to make the plant a much more hazardous place than usual. A 'safe system of work' must therefore be implemented – to safeguard personnel, property and the environment – and should consist of five major elements:

> *The safety team.* Led by a safety officer, the team is responsible for a number of safety functions including, but not limited to:

> - developing a safety strategy for the turnaround;
> - briefing all personnel on safety before the event commences;
> - providing help, advice and assistance on safety to all personnel during the event;
> - providing and controlling guardians for tasks requiring entry permits;
> - co-ordinating emergency marshals and controllers.

It can be seen from Figure 9.14 that the safety communications network is complex and needs to be well defined and operated effectively.

Safe working routine. This should contain safety guidelines for ensuring that the workplace and surrounding environment are safe, materials used are not hazardous and that people are competent and well briefed on the requirements and hazards of the tasks they are required to perform. The routine should be driven on a day-to-day basis by the supervisors.

Task hazard assessment. A process for analysing and dealing with the hazards involved in performing a specific task, this has four stages:

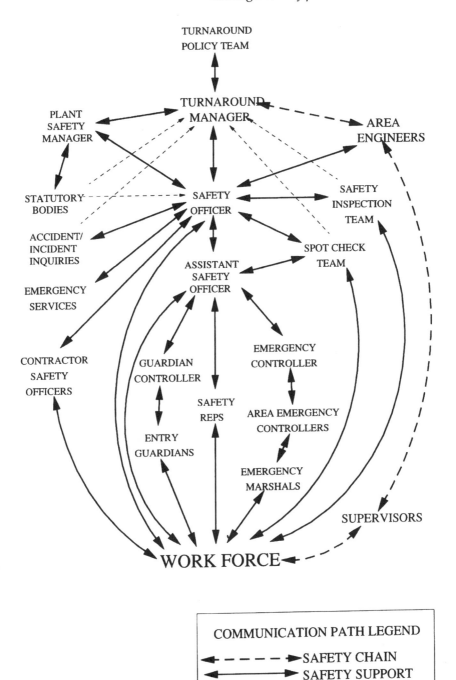

Figure 9.14 The safety communications network

(i)   defining the main steps of the task;
(ii)   for each step, identifying any hazards involved;
(iii) specifying what type of loss is attached to each hazard;
(iv)   specifying action to either eliminate or guard against the hazard.

The assessment should be carried out on all major tasks and on a selection of minor and bulkwork tasks. Also, if any member of the team should have reservations about any task it is essential that it be subjected to a hazard assessment.

Safety inspections. The working site must be continuously monitored in order to ensure that the safe system of work is adequate and that personnel comply with it. Two formal inspection routines are vital elements in the monitoring programme. The first is general safety inspection carried out on a daily basis by a team of managers who should be invited by the safety officer to take part; they should look for unsafe acts, unsafe conditions and instances of untidiness and bad housekeeping around the site in general. The second is recurrent spot checks carried out on randomly selected tasks by a team who should be invited by the safety officer to answer the following three questions.

> *Are the safety measures specified for the tasks adequate?*
> *Are they understood by those performing the task?*
> *Are they being complied with?*

Information gathered via both routines should be fed back, for possible action, to the daily turnaround control meeting.

Accident investigation. If, in spite of the safety measures taken, an accident should occur, it should be the responsibility of the turnaround manager to convene an investigation to ascertain:

- the type and extent of loss inflicted on people, property or the environment;
- the nature of the specific incident that caused the loss;
- the immediate reasons for the incident;
- the root causes underlying the accident.

Formulation of the quality plan

Every task should be properly specified, and executed to specification. Critical tasks should be checked, on completion, by qualified and plant-based quality teams to ensure that the work has complied with plant quality standards. All tasks should be signed off by a plant-based person. The quality plan should define the critical tasks and identify the plant-based personnel who have the authority to sign-off tasks.

Briefing of all turnaround personnel

The purposes of briefing are to:

- provide accurate general information;
- alert everyone to the rules governing the turnaround;
- create a common understanding among – and to gain commitment from – all those involved.

Typically, it will provide information about:

- timing, duration and work patterns;
- local manpower and contractors;
- workscope and schedule;
- the turnaround organization;
- accommodation and facilities;
- the safety system of work and the quality plan.

Phase 3: Executing the turnaround

Once preparation is complete and all personnel have been briefed, the turnaround manager should take charge of the event. Figure 9.15 outlines the turnaround manager's daily routine.

The shutdown of the plant

The shutdown of the plant should normally be controlled by the plant manager and the team. The turnaround manager should supply the resources to perform the civil, mechanical, electrical and instrumentation work of the shutdown. From 'product off' the plant would be taken through run-down of stock, cleaning, cooling, sweetening of the atmosphere inside vessels and equipment, and isolation of all equipment – to the point when the safety team would carry out atmospheric tests in vessels and equipment to ensure the absence of noxious, toxic or volatile substances. The plant could then be said to be 'dead', i.e. safe to work on.

Carrying out the turnaround tasks

The first few days of this phase should be devoted to opening up large items of equipment (vessels, columns, rotating machinery, etc.) and stripping out small items (valves, small pumps, motors, etc.). After stripdown the main activities would be plant inspection, repair, refurbishment and equipment cleaning. At this point large specialized tasks, such as catalyst renewal, column re-traying and compressor overhaul would be started.

- CHECK PREVIOUS 24 HOURS PROGRESS WITH THE PLANNING OFFICER
- CHECK COST CONTROL AND FORECAST WITH THE QUANTITY SURVEYOR
- VISIT THE SAFETY CABIN AND CHECK ON SAFETY ISSUES
- VISIT THE STORES AND CHECK ON DELIVERY AND ISSUE PROBLEMS
- VISIT THE WORKSHOPS AND CHECK ON DAILY PROGRESS
- TOUR THE SITE TO CHECK ON SAFETY & HOUSEKEEPING - TALK TO PEOPLE
- TAKE PART IN ANY SCHEDULED SAFETY INSPECTION OR SPOT CHECK
- VISIT PERMIT TO WORK ISSUERS TO DISCUSS ANY ISSUES
- VISIT THE QUALITY TEAM AND DISCUSS QUALITY ISSUES
- VET OVERTIME REQUESTS AND APPROVE/ MODIFY REJECT

- MEET WITH THE PLANT AND ENGINEERING MANAGER
   ~ RESOLVE TECHNICAL PROBLEMS
   ~ DISCUSS & APPROVE/ REJECT REQUEST FOR EXTRA/ ADDITIONAL WORK
   ~ FORMULATE STRATEGIES TO KEEP THE TURNAROUND ON PROGRAMME
   ~ RESOLVE INDUSTRIAL RELATIONS PROBLEMS
   ~ RESOLVE INTERFACE CONFLICTS
   ~ DEFINE CONSEQUENCES OF ANY CHANGE OF INTENT

- CHAIR THE TURNAROUND CONTROL MEETING
   ~ SAFETY OFFICER REPORTS ON SAFETY ISSUES AND ANY INCIDENTS
   ~ AREA ENGINEERS REPORT ON AREA WORK PROGRESS AND PROBLEMS
   ~ PROJECT MANAGERS REPORT ON PROGRESS OF PROJECTS
   ~ PLANT MANAGER REPORTS ON ANY PLANT ISSUES
   ~ MAINTENANCE MANAGER REPORTS ON ANY ENGINEERING ISSUE
   ~ QUALITY TEAM LEADER REPORTS ON QUALITY ISSUES
   ~ QUANTITY SURVEYOR REPORTS ON EXPENDITURE AND COST ISSUES
   ~ CHAIRMAN SUMS UP / ISSUES INSTRUCTIONS/ DELEGATES TASKS

- WRITE A DAILY TURNAROUND REPORT AND ISSUE IT

Figure 9.15 Turnaround manager's daily routine

Defining and costing the extra work

Inspection of equipment will often reveal faults which were not predicted and which require the carrying out of work which is *extra* to that which has been planned (but note that inspections undertaken during previous shutdowns, inspections undertaken on-line, and effective pre-planning should minimize the occurrence of this). Such work should be specified, costed and submitted for approval to the plant and turnaround managers on a daily basis. If the work required is going to have a negative impact on any of the turnaround objectives it should be submitted to the policy team for action. Should the work be approved it should be entered on an extra worklist and added to the work schedule. Occasionally a fault could be revealed which could have such a serious impact on turnaround objectives that it could require decisions to be made by management at the very highest level.

Defining and costing additional work

During the event, work may be exposed which was not on the worklist because it was either not considered necessary or was overlooked by the plant team. This *additional* work should be treated in exactly the same way as extra work except that it should be recorded on a separate worklist. After the turnaround this should be investigated to ascertain the reasons why it was not included in the original worklist.

Monitoring progress, productivity, safety, quality and expenditure

The main monitoring vehicle should be the daily control meeting. Figure 9.16 (a) is a comprehensive checklist of the subject matter that should be covered. This formal tool should be supplemented by a process of continuous communication between members of the team. A five or ten-day window of the kind outlined in Figure 9.16 (b) is often adopted for monitoring the progress of the turnaround.

Starting-up the plant

A point will eventually be reached when most of the tasks will either have been completed or be nearing completion and the decision could be taken to disband the turnaround organization and replace it with a 'start-up' team. This should be led by the plant manager with the turnaround manager assuming a support role. The daily control meetings should be replaced by regular start-up meetings (often held twice a day). The start-up phase will be a mixture of completing any remaining turnaround tasks and bringing plant systems back on line.

The daily control meeting must provide the following information and actions

SAFETY OFFICER'S REPORT	AREA ENGINEERS' REPORTS

~ details of accidents/ incidents in last 24 hours
~ details of any recurring accidents/ incidents
~ findings of daily site inspection and spot check
~ summary of site safety level and details of any
    particular safety concerns
~ recommendations for safety improvements
~ details of any safety initiatives or awards
~ tomorrow's safety slogan
	~ progress on major tasks including any technical
	    problems and solutions
	~ tasks completed, boxed up and handed back and
	    percentage completion of other major tasks
	~ progress on small tasks and bulkwork
	~ any hold-ups or shortages on manpower, materials,
	    equipment or services
	~ any conflicts with other areas of work
	~ whether the area is on schedule or behind - and,
	    if behind, the strategy for getting back on target
	~ assessment of unavoidable overrun, how many
	    hours or days, and why it is unavoidable

PROJECT MANAGER'S REPORT	PLANT MANAGER'S REPORT

~ progress to date on project including any technical
    problems and solutions
~ progress on any "break ins"
~ any "bad fit" problems due to poor design
~ any hold ups or shortages on manpower, materials,
    equipment or services
~ any conflicts with other areas of work
~ whether the area is on schedule or behind - and,
    if behind, the strategy for getting back on target
~ assessment of unavoidable overrun, how many
    hours or days, and why it is unavoidable
	~ current ability of permit to work issuers to issue
	    permits on time & strategy to eliminate any delays
	~ any handover quality issues
	~ general view of on site performance
	~ general view of on-site housekeeping
	~any upcoming on-site problems
	~ warning of any system coming back on line early
	~ warning of any process activity that could impact
	    turnaround progress or safety

MAINTENANCE MANAGER'S REPORT ~ any concerns on turnaround progress ~ any engineering concerns ~ any quality performance concerns ~ any upcoming engineering problems ~ any questions on turnaround engineering work being done - QUANTITY SURVEYOR'S REPORT ~ actual expenditure to date vs planned expenditure ~ expenditure trends in each area ~ specific examples of cost overrun ~ general forecast on final turnaround cost ~ any recommendations for tighter cost control or cost saving initiatives	QUALITY TEAM LEADER'S REPORT ~ quality trends in the last 24 hours ~ any specific quality problems ~ any recurring quality problems ~ any recommendations for quality improvement

Figure 9.16 (a)   The daily control meeting

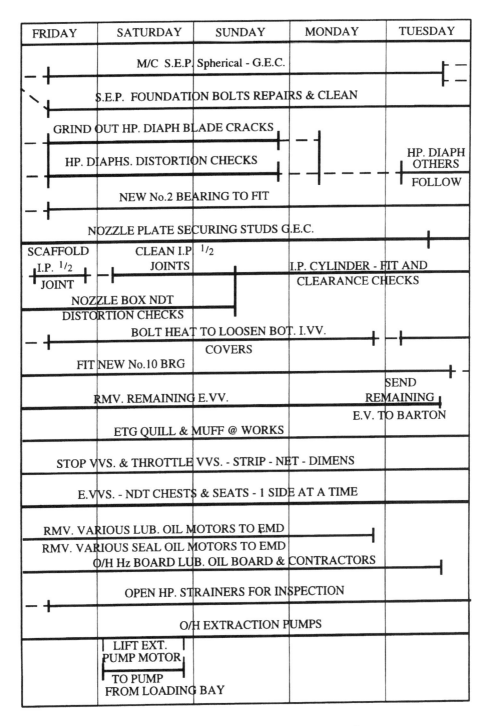

Figure 9.16 (b)   Bar chart for a five-day window for shutdown control

~ open the meeting and control through the chair

~ ask for and note reports in pre-set order

~ ask specific questions to clarify points

~ do not allow detailed discussion of issues at this
    meeting - convene separate discussions

~ sum up general progress on key indicators

~ voice any concerns on trends or specific issues

~ make executive decisions and inform the meeting of
    them, their requirements and consequences

~ delegate specific actions to particular people

~ delegate responsibilities to convene further discussions
    on key issues outside of the control meeting

~ announce next day's quality initiative

~ announce next day's safety slogan

~ make any other announcements

~ state, and ask for, any other business

~ close the meeting

Figure 9.16 (c)   Turnaround manager's control meeting routine

Phase 4: Terminating the turnaround

During the start-up phase, and for approximately four weeks after, actions should be taken to demobilize all turnaround resources and to return the plant area to its former state. In addition to this, a formal de-brief should be conducted while events are still fresh in people's minds, to record what happened and any lessons to be learned. The final action of the termination phase should be for the turnaround manager to compile and issue a report which should detail the work done, compare actual against planned performance and make recommendations for future events of the same kind.

10
Maintenance management control

Introduction

Chapter 1 emphasized the importance of establishing a maintenance objective – it is the starting point of the strategic management process. The linkage between that objective and the process of maintenance control was shown in Figure 1.1 – our maintenance management paradigm – where it was indicated that the control system is needed to direct the maintenance effort towards the objective.

In the case of the alumina refinery example of Figures 4.1 and 4.2 the important control questions are:

- Is the maintenance effort achieving the desired availability levels of 95%?
- Is the incurred maintenance cost within budget?
- If the answer to either or both of these questions is no – what are the reasons?

Information feedback of this kind allows the maintenance effort to be adjusted and/or re-directed as necessary.

Principles of maintenance control

The key relationships and processes of maintenance management control are delineated in Figure 10.1, a model based on the concepts outlined in Chapter 1. The strategy (life plans, organizational policy, etc.) is established by maintenance management in order to achieve the objective (the agreed plant-user requirement at the anticipated cost). Management budgets for – and uses – resources (men, spares, tools) to implement the strategy.

The reporting system has the following principal data collection functions:

- checking whether the maintenance strategy is being carried out to specification (Point 1);
- measuring the parameters of maintenance output (Point 2);
- measuring the maintenance resource costs (Point 3).

The cause of any detected deviation from intent can then be determined and the necessary corrective action taken.

The above would seem to be a straightforward enough process. In practice,

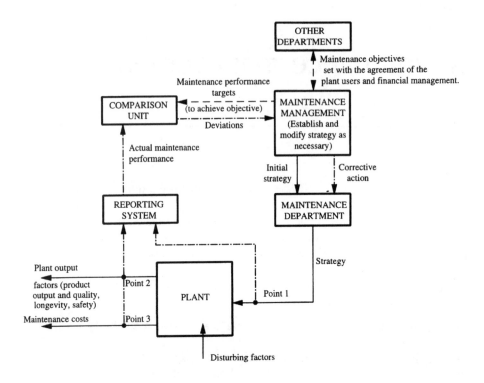

Figure 10.1 Maintenance management control

however, there are many complicating factors:

- A deterioration in some measured output could have causes other than maintenance – maloperation – for example. So the root causes of the deviation must be pin-pointed before any control measure is taken.
- Once set, maintenance objectives may not be unchanging; they will often have to be amended in the light of new needs of other departments, and before the overall strategy has had time to 'take effect'.
- Although the direct maintenance costs are relatively easy to measure some of the parameters of maintenance output, such as longevity or safety, are not.
- It is frequently the case that requirements relating to product output or quality will vary in the short and medium term, those relating to plant longevity and safety in the much longer term. Indeed, the maintenance strategic effort as regards plant longevity and safety is often quite divorced from that regarding product output and quality.
- The direct maintenance costs are a function both of maintenance policy (see Figures 10.2 and 10.3) and of organizational efficiency

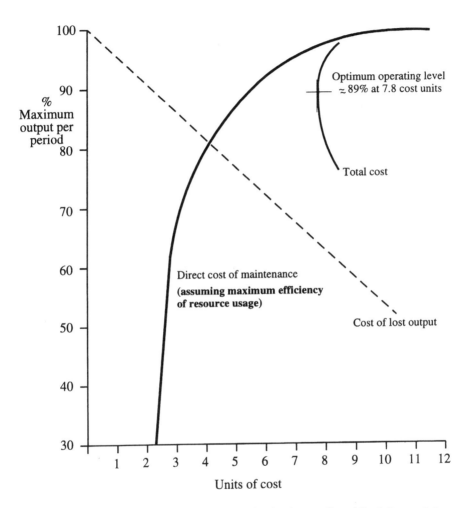

Figure 10.2 Establishing the optimum operating level using a policy of fixed-time maintenance

(see Figure 10.4). These two aspects may well need separate objectives and control systems.

For these reasons, any overall maintenance control system based on the model of Figure 10.1 will have its limitations. Although it will be possible to identify deviations – from targets for output parameters and for maintenance resource costs – it will be very much more difficult to identify the causes. For this to be possible it may well be necessary to have a hierarchy of objectives and corresponding control systems (see Figure 10.5). (Maintenance

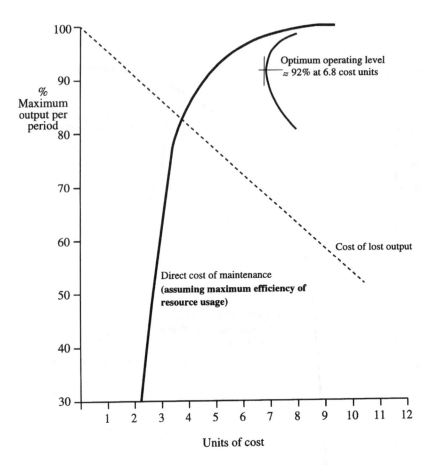

Figure 10.3 Establishing the optimum operating level using a policy of condition-based maintenance

objectives were discussed in Chapter 7 of Book I – from which Figures 10.2 to 10.5 are taken.)

Practical interpretation of maintenance control principles

The best practical mechanism for controlling the overall maintenance effort (Maintenance Productivity) would be a properly designed maintenance costing system incorporating the ideas of zero-based budgeting. Thus, for the alumina refinery of Figure 4.1, the strategic life plan and its budget have been set up to achieve the plant maintenance objective (the user-requirements at budgeted cost). A maintenance costing system (see Figure 10.6) could be designed which

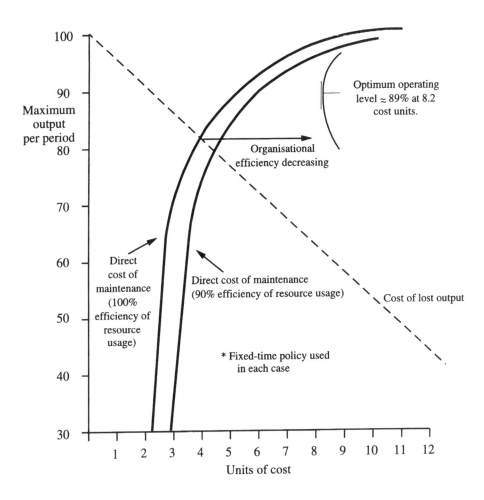

Figure 10.4 Effect of efficiency of resource usage on the location of the optimum operating level

would facilitate determination – for each unit of plant (for a bauxite mill, say), and therefore for the refinery – of the relationships between the rates of expenditure on maintenance and the relevant output parameters. The maintenance costs could be categorized by resource type (men, spares, etc.), maintenance type (preventive, corrective, etc.), trade (mechanical, electrical, etc.), budget type (short term for, say, availability or quality; long term for plant longevity). In addition, costs could be assigned against supervisor, trade group or major job (i.e. an overhaul). Traditionally, such a system relies on coding the plant units, work groups, work types, etc. (see Figure 10.6). All work is therefore recorded on work order (or time) cards – against unit, type, and work group – and all spares usage similarly recorded on stores requisition orders, etc.

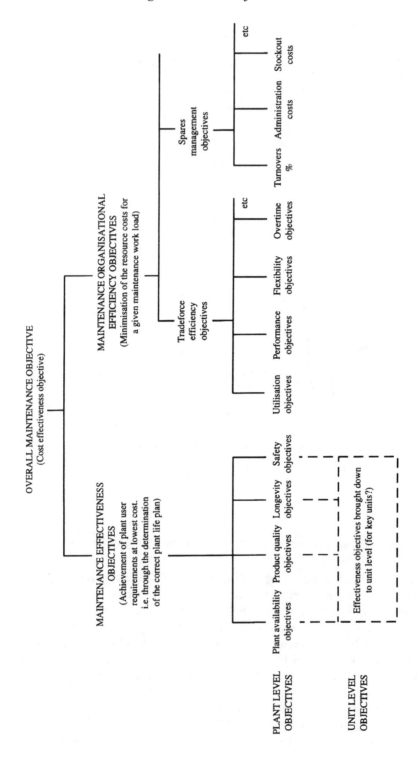

Figure 10.5 Hierarchy of maintenance objectives

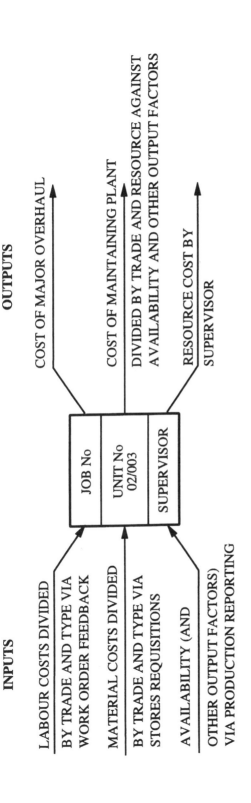

INPUTS

LABOUR COSTS DIVIDED
BY TRADE AND TYPE VIA
WORK ORDER FEEDBACK

MATERIAL COSTS DIVIDED
BY TRADE AND TYPE VIA
STORES REQUISITIONS

AVAILABILITY (AND
OTHER OUTPUT FACTORS)
VIA PRODUCTION REPORTING

OUTPUTS

COST OF MAJOR OVERHAUL

COST OF MAINTAINING PLANT
DIVIDED BY TRADE AND RESOURCE AGAINST
AVAILABILITY AND OTHER OUTPUT FACTORS

RESOURCE COST BY
SUPERVISOR

| JOB No |
| UNIT No 02/003 |
| SUPERVISOR |

Job No.	Plant code		Trade and supervisor		Work type
	Plant	Unit	Electrician	Night shift	Preventive
521	02	003	2	NS	2

Figure 10.6 Outline of a maintenance costing system

The system can be designed to provide a variety of outputs, either automatically or on demand – especially if the processing is computerized. The main outputs and their possible uses are as follows:

(i)   Actual maintenance costs (separated into labour and material costs and, if required, divided according to work type and trade) and recorded levels of relevant output parameters (availability, product quality losses) – which can be compared against budgeted costs and targeted levels of performance (**per period, per unit and per plant**).

(ii)   Identified areas (plants or units) of high maintenance cost or low availability, perhaps presented via Pareto plots or 'top ten' ranking lists).

(iii)   Plots of output performance versus maintenance costs, per unit or per plant.

(iv)   Actual maintenance costs per trade group or per supervisor – for comparison with budget.

(v)   Actual cost of major overhauls – again for comparison with budget.

The above, although not ideal, should satisfy most of the requirements of an overall maintenance control system, facilitating:

- the setting of objectives,
- the monitoring of output parameters (such as availability) and of inputs (such as resource cost relative to budget) which can influence the levels achieved,
- the diagnosis of deviations from intent and the prescription of appropriate remedial action.

Most industrial budgeting and costing systems are designed by accountants for corporate financial control and are not sufficiently equipment-oriented to shed light on the problems of maintenance control, e.g. cost centres may not be plant-specific and, even when they are, each one may encompass too large an area of plant to be of any use in maintenance management. In addition, only rarely do such systems have the facility of comparing maintenance costs against the various parameters of output.

Even if properly designed, a maintenance costing system has to be a high level, longer-term one, providing a means of controlling the *overall* maintenance effort. It will be appreciated from Figure 10.5 that it needs to be complemented by control systems operating at a lower level (and on a shorter time scale). Indeed, it could be argued that a control system is needed for each objective that is set. For example, if an overtime limit is set then the actual overtime needs to be monitored and reported to the supervisor for necessary corrective action (see also Figure 10.5). In other words, control systems are integral to the operation of organizations.

The two principal lower level maintenance control systems (see Figure 10.5) are those which deal with *effectiveness* and *organizational efficiency*, respectively. The former is concerned with ensuring the effectiveness of the plant maintenance life plan as regards achieving desired outputs and meeting cost targets, the latter with ensuring that maintenance work is being carried out in the most efficient way.

The control of maintenance effectiveness

This is perhaps *the* most important maintenance control system (it is also discussed in Chapter 10 of Book I). Once again, the alumina refinery will serve as the vehicle for explaining its operation (see Figure 4.1).

Figure 10.7, which outlines the mechanisms for controlling the effectiveness of one of the refinery units, illustrates the classic ideas of *reactive* control – using the feedback of operational and maintenance data – and also highlights *pro-active* control via the feedforward of ideas for reliability and maintenance improvement.

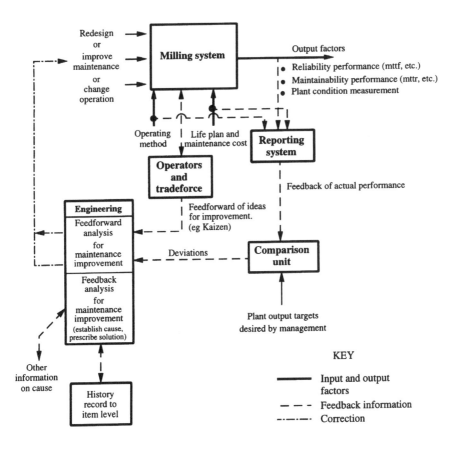

Figure 10.7 Controlling the reliability of a unit of plant

Reactive control of plant reliability. The requirements of the systems are to:

(a)   monitor the output parameters of each unit, e.g. reliability (mttf), maintainability (mttr), plant condition, etc. and some of the input conditions, e.g. whether the unit life plan is being carried out to specification and at anticipated cost;

(b)   determine the root cause of any failure (Note that a control system for this must encompass several departments because the cause could originate in Production (maloperation), in Engineering (poor design) or in Maintenance.);

(c)   prescribe the necessary corrective action.

At refinery level, control can be envisaged as in Figure 10.8, i.e. each unit having its own control system. Once again, the difficulty is caused by the multiplicity of units which make up a major industrial plant – and therefore of control systems needed. The consequent data processing has been made manageable by modern computer technology which can easily handle the many independent control mechanisms. The difficulty, however, usually lies not in the processing but in the acquisition of the data. Company management may therefore need to concentrate control effort on selected units, those which they deem critical; for the rest they may use the maintenance costing system to identify the most troublesome, e.g. those of highest high maintenance cost, poorest product quality, highest downtime, and so on.

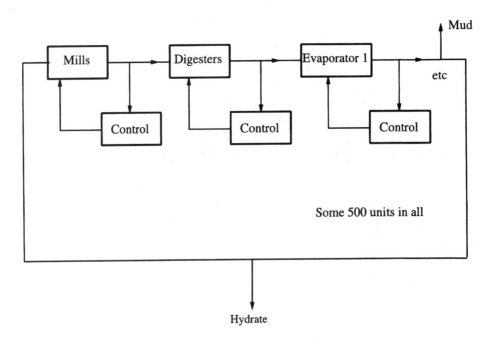

Figure 10.8 Controlling the reliability of a plant

Pro-active control of unit reliability. Figure 10.7 also illustrates the pro-active approach, which differs from the reactive in that it does not wait for failures or for high cost problems to occur before taking action. The basic idea is that all members of the organization – but especially the shop floor – should continuously seek ways of improving unit reliability, and hence output, safety, and so forth. The Japanese call this *kaizen*. The shop floor form small inter-disciplinary – but plant-orientated – teams (see Figure 10.9) to improve the reliability of selected units. (Preventive maintenance is interpreted literally – to prevent the need for *any* maintenance, by design-out and other actions). Upper management circles ensure that the idea is promoted and accepted throughout the organization. This ensures that middle management circles give assistance and advice to the shop floor teams as necessary.

Incorporating reliability control systems into the organization. Although Figures 10.7 and 10.8 are useful for understanding the mechanisms of reliability control it still remains to incorporate these ideas into a scheme for a working maintenance organization. This is shown in the general model, Figure 10.10, and the application of this to the alumina refinery organization, which is outlined in Figure 10.11.

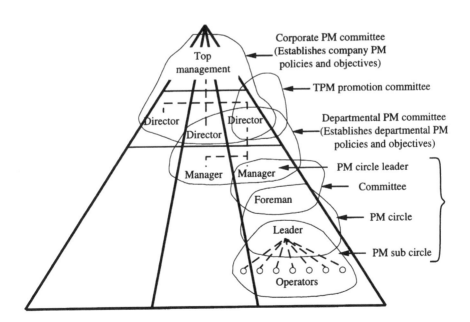

Figure 10.9 A system for promoting TPM within an existing organization

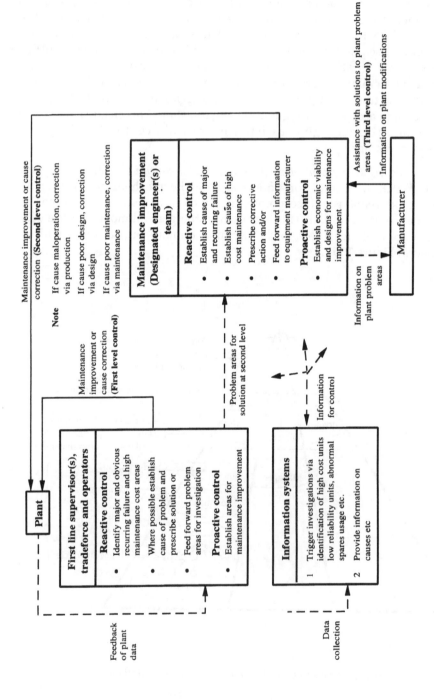

Figure 10.10 General model of reliability control within an organization

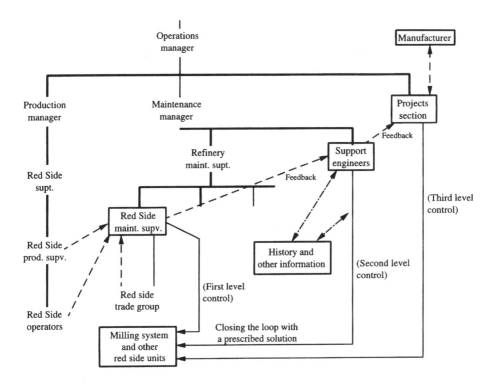

Figure 10.11 Plant reliability control within the refinery organization

It can be seen that there are several interrelated levels of plant reliability control in an organization – each with its own responsibilities and roles. The first operates between the shop floor and supervisors and to a large extent is independent of the information systems – however, a history record can be important here. This level of control is particularly useful because there is a quicker reaction to problems. Because the personnel involved may be present at a repair, and can discuss it with operators and tradesmen, there is likely to be first hand knowledge of the cause of failure. In addition it is at this level that the main thrust of pro-active control operates; if the personnel involved cannot establish the cause and/or prescribe and implement a solution, then the problem is passed up to the second level.

The second level of control operates through designated engineers and/or a maintenance investigation team. To be effective, this requires the integration of information systems and engineering investigation. The information system (computerized) should be designed around the ideas illustrated in Figure 10.12 and should therefore be capable of identifying problem units and hence triggering corrective investigation within the organization (satisfying Point (a) of the control requirements). The major effort, however, will lie in

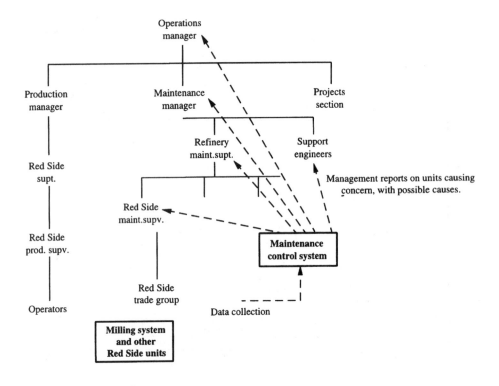

Figure 10.12 The use of maintenance information systems for reliability control

the diagnosis of failure causes (Point (b)), by interrogating the plant history, and in the prescription of corrective action (Point (c)), an effort which will need to come from the investigative engineers.

In general it is the root cause of any problem which will be sought, and because investigative effort is necessarily limited only a small number of problem items can be looked into at any one time. The criterion for selecting these is usually based on some kind of item ranking, by downtime, direct cost or failure frequency (see, for example, Figures 10.13(a) and (b) which show such rankings, Pareto analyses, for a mining vehicle fleet[1]).

A possible third level of control lies in the contact between the various users of a given type of equipment and its manufacturer. This offers the opportunity for maintenance information to be collected from a much larger pool of experience. However, because more than one user company will, in practice, be involved it is the least effective level of control. The onus for ensuring the success of such an activity rests with the equipment manufacturer.

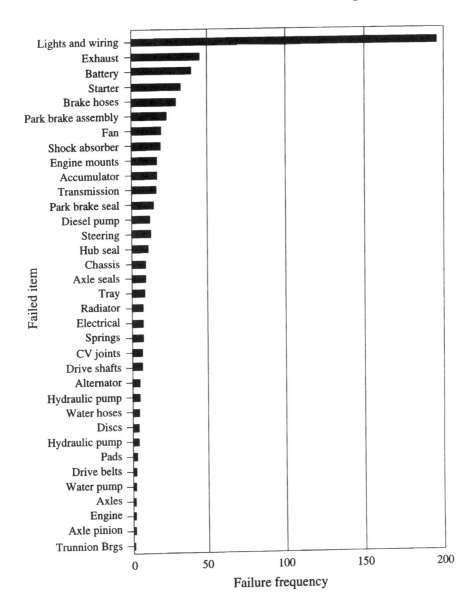

Figure 10.13 Pareto analyses, mining vehicles: (a) Failure frequency

The control of organizational efficiency

The concept of organizational efficiency (see Figure 10.4) was discussed in Chapter 7 of Book I where it was stated that the prime organizational objective is:

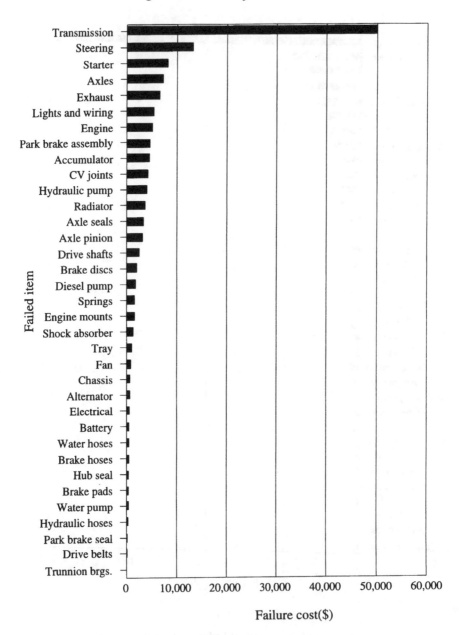

Figures 10.13 Pareto analyses, mining vehicles: (b) Failure cost

to carry out a given plant maintenance life plan at minimum cost by using maintenance resources (man, spares, tools) in the most efficient way.

However, it was also pointed out that a single objective for organizational efficiency is somewhat fanciful. A hierarchy of sub-objectives was therefore developed, as shown in Figure 10.5.

For tradeforce efficiency the objective might be stated as:

> to minimize the tradeforce cost per period for carrying out a given plant life plan.

The extent to which this was being achieved could be measured via labour efficiency indices such as :

Tradeforce performance   = $\dfrac{\text{Standard hours in workload}}{\text{Actual hours paid}}$

Tradeforce utilization   =   Percentage of time active per period

For this, one or more of the techniques of maintenance work measurement and/or work sampling would have to be used[2]. Work measurement is expensive and can bring industrial relations problems in its wake, but control that is based on it is superior to control based solely on cost objectives because it can measure labour efficiency against universal standards. Other things being equal, it is in the interest of maintenance management to pursue organizational actions – encouraging inter-trade flexibility, accelerating the introduction of self-empowered teams, making more use of contract labour, etc. – that will lead to improvements in the indices obtained from such efficiency measurements.

In the case of spare parts the objective is:

> to minimize the sum of stock-out costs and holding costs.

An appropriate index of stores efficiency might therefore be one which involved holding costs versus number or cost of stockouts. Other sub-measures of stores efficiency could be derived via the monitoring of such parameters as percentage stock turnover, average tradeforce waiting time for parts, staffing costs versus stockholding costs (see Chapter 11).

It must be accepted that there are considerable practical difficulties in measuring and using such indices. An alternative approach, which could be applied to simple first and second line jobs (different information would be needed for third line, major overhaul, work), is illustrated in Figure 10.14. This could provide simple indices, e.g. a job-delay ratio would give a measure of organizational efficiency. In addition, profiling information on the possible causes could point to appropriate corrective action.

Published methods for measuring maintenance performance

Various index-based methods for measuring maintenance performance, and hence for controlling maintenance effort, have been developed[3] but examples of industrial organizations which have used them are hard to find. Exceptions to this are the Japanese methods that have been incorporated into their concept of Total Productive Maintenance[4].

Was there a delay in carrying out the job?

If yes how long?

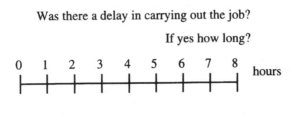

Did the delay cause production loss?

If yes how long?

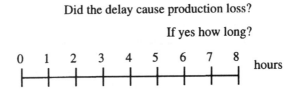

What was the main cause of the delay?

Stores	Equipment availability	Tools	Isolation	Other trades	Job instructions

Figure 10.14   Collecting data on organizational performance

The use of indices of maintenance performance for inter-firm comparisons

A different use of indices is for comparing the maintenance departments of various companies. It is within such a context that the objectives developed in Figure 10.5 could be extended and developed – into a hierarchy of maintenance performance indices (see Figure 10.15). The idea is that a series of indices categorized as shown (many of which may be used for control purposes) could be profiled in order to compare the maintenance performance between plants of similar technology and size – alumina refineries, say. The value of doing this is not obvious, however. Even when comparing alumina refineries there are many differences – in detailed design, size, technology, manufacturer, etc. – that exercise greater influence on the indices than do such aspects as the maintenance life plan or organization.

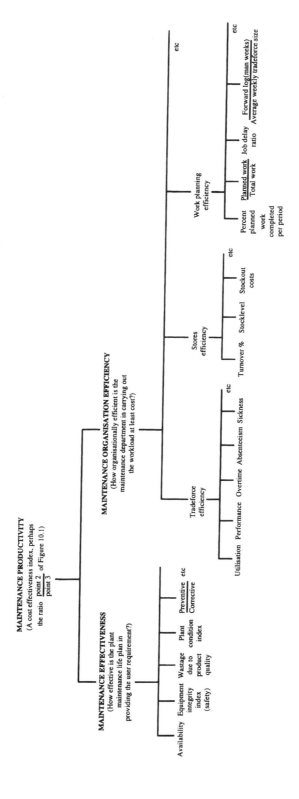

Figure 10.15   Proposed hierarchy of maintenance performance indices

First line maintenance
Fitter removes item and returns
to workshop for repair

Workshop fitter repairs
before returning to stores

ASK FOR

(1) Date (running hours)

(2) Symptoms

(3) Probable cause of failure

ASK FOR

(1) Component(s) that caused
loss of function

(2) Probable cause of failure
(as before)

	NWT	Maloperation	Poor design	Poor maintenance
Tick				

Plus comment on why box ticked

Figure 10.16   Collecting data on the cause of failure

Table 10.1 Factors conducive to good data collection

Senior managers	Use an open approach Sell the idea Introduce the system gradually Do not use the system to assess the data inputter Show commitment to the system
Foremen	Must be committed to the system Must be convinced the system will benefit them
Tradesmen	Handle simple paperwork or user-friendly software Are allowed ample time for information input between jobs Have access to the system Appreciate what the data is used for
System designers	Make effective use of the data and ensure that this is seen to be the case Collect data in the easiest manner Limit the data collected to only that which is needed

Human factors and maintenance control

The success of pro-active control, one of the cornerstones of Total Productive Maintenance, depends on the quality, goodwill and motivation of the shop floor; that of reactive on the quality of data returned from that level. With regard to the latter, the author's own studies have shown that data collection systems rarely operate well and that human factors problems and lack of training feature as the most important reasons for this (see Table 10.1). Many of the companies investigated had sophisticated computer systems for control but had put little effort into defining terms such as symptom, root cause of failure, defective part, etc. It is not surprising, therefore, that the quality of data feedback was poor – especially concerning the cause of any problem. Perhaps the alternative means of data collection on cause, shown in Figure 10.16, would help in overcoming this problem.

References

1. Healy, A., *Effect of road roughness on the maintenance costs of 4WDs*. PhD thesis, Queensland University of Technology, Brisbane, Australia 1996.
2. Kelly, A., *Maintenance Planning and Control*, Appendix 2. Butterworths 1984.
3. Jardine, A. K. S. (ed.), *Operational Research in Maintenance*. Manchester University Press 1970.
4. Hibi, S., *How to Measure Maintenance Performance*. Asian Productivity Association 1977.

11
Maintenance stores

Introduction

The theory and practice of stores management is a subject in its own right, quite a few textbooks having been solely devoted to stores management in general and stores inventory control in particular, and dealing with applications in a wide range of industrial, commercial, wholesale and retail organizations. Although some of the basic aspects of the management of maintenance spares are covered by this material there are some unique areas of the problem that are not – the management of high-cost, slow-moving, spares, for example. The objective of this chapter is therefore to provide an introduction to this rather specialized activity, the operation and management of the maintenance spare parts storage system.

System function and objective

The operation of the maintenance stores system is modelled in Figure 11.1. The basic function of the stores is to act as a buffer (or reservoir) between the delays and uncertainties of the supply from the manufacturer (or from the reconditioning workshop) and the inherent variability of maintenance demand – and thus reduce plant downtime caused by waiting for spares. In practice, however, there is a limit to the extent to which this aim can be pursued. If large numbers of every conceivable spare were always held then such downtime would be minimal, but the costs of obtaining and holding stock would be excessive. The rational objective, therefore, of running a spares stores and controlling its inventory is:

> to minimize the sum of the associated direct costs (of obtaining and holding the spares) and indirect costs (of loss of production due to waiting for spares).

In this way the maintenance stores will make its greatest possible contribution to minimizing business costs.

Outline of system operation

For ease of identification and retrieval maintenance parts are coded by type and catalogued. Each part type requires an inventory policy, a set of rules for deciding how the number held in store is to be controlled so that the stores objective will be met. One of the most commonly used, for example, is for

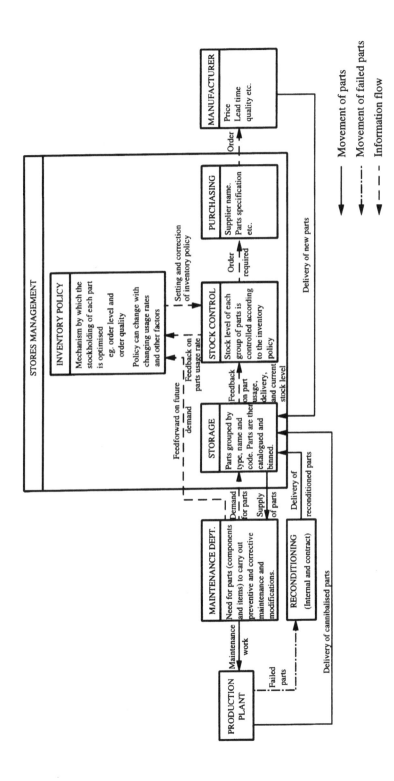

Figure 11.1 Maintenance spares management system

fast moving parts, and is based on the notion of a cost-optimal *re-order level* (for any particular part, the stores holding which, when reached, will trigger the process of re-ordering) and a cost-optimal *economic order quantity* (the number to be ordered at any one time). Thus, stock control monitors the usage anddelivery of parts and uses the inventory policy to control replenishment.

Note that the inventory policy depends on such parameters as demand rate and lead time. It is therefore important that they are monitored both via data feedback (on actual recorded usage) and via information feedforward (on anticipated changes in maintenance life plan or production usage) to ensure that any changes are identified and the inventory policy modified accordingly.

If there were only one type of spare in stores and the demand for it was high then stores management would be easy – although Figure 11.1 would still describe the operation of the system. In practice, however, the main difficulties are:

(a) *Multiplicity of parts and material types.* Stores in even a small plant might well hold over a thousand different types of item – hence the need for cataloguing. The onus should be on the engineering and maintenance departments to reduce this multiplication via procurement policies based on rationalization and interchangeability. A complicating factor, however, is that in industrial companies the maintenance requirement accounts for only a part of the total stockholding; there will also be production stock, commercial stock, and so on. The question then often arises as to whether maintenance should manage its own stores or whether stores management should be centralized under the commercial department (see later).

(b) *High total cost of the spare parts holding.* A medium-sized power station might well hold stock to the value of sixteen million pounds (1996 values), a small company to the value of five per cent of the capital replacement value. A typical analysis of such a stock-holding often reveals that some 80% of the total cost is accounted for by 20% of the items held (see Figure 11.2) More often than not these high cost items tend to be the slow or very slow moving spares.

(c) *Wide range of usage rates and lead times.* Each of the part types requires an inventory policy which depends largely on usage rate and to a lesser extent on lead time. As explained, there are various well-validated quantitative techniques for determining the inventory policy for fast moving parts. There are relatively few, however, when it comes to slow movers – and it is in this area that high costs are incurred.

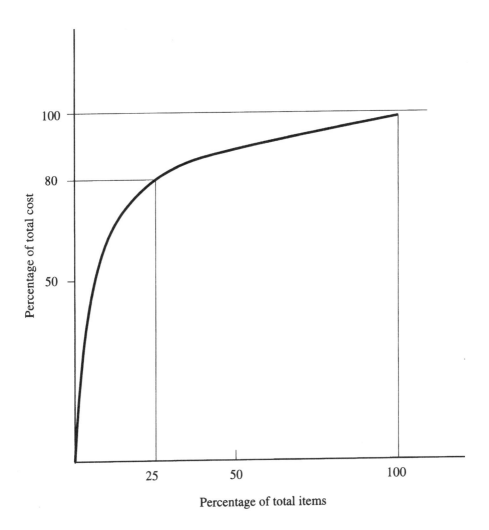

Figure 11.2   Pareto plot of probable spares distribution

Inventory policies for the control of fast moving spares

To facilitate the setting of inventory control policies spare parts can be classified according to their usage rates into *fast moving* (where the demand is greater than, say, three items per year) and *slow moving* (demand less than that figure).

Inventory control policies for fast moving spares have been covered fairly extensively in various textbooks (including the author's)[1]. Just one such policy will therefore be presented here, and then only in sufficient outline to illustrate the basic principles, the main effort of this chapter being directed at the problem of slow moving spares.

As already explained, the task is to balance the cost of holding stock against the cost of running out. In general, inventory control theory attempts to determine those procedures which will minimize the sum of the cost of:

- *running out* of stock (production loss due to stoppage, cost of temporary hire, etc.),
- *replenishing* stock (which in part depends on the quantity ordered),
- *holding* stock (interest on capital, depreciation, insurance, etc.).

There are two basic categories of control policy for fast movers, namely:

- *Re-order level*: replenishment prompted by stock falling to a pre-set re-order level;
- *Re-order cycle*: stock reviewed, and replenishment decided, at regular intervals.

Lewis, in his concise and readable textbook[2] on this topic, further sub-divides these categories into five policy types (a classification which, he states, is exhaustive, other types usually being further elaborations on one or more of these basic schemes). For illustration this section will describe a re-order level policy.

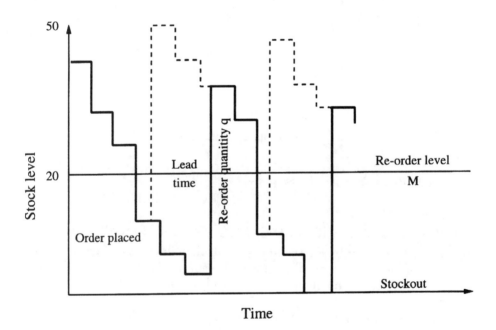

Figure 11.3   Variation of stock level under a re-order level regime

A re-order level policy – the so-called 'two-bin' system (see Figures 11.1 and 11.3)

The inventory policy is set in terms of a re-order level M and a re-order quantity q. The stock is continuously monitored and a replenishment order for a fixed quantity q is placed when stock on-hand (stock held plus stock on order) falls to or below a pre-set re-order level M (i.e. storage in two bins, order placed when first bin empty, service from second bin until order received). The re-order level stock (i.e. the contents of the second bin) thus acts as a reservoir which diminishes the risk of running out of stock arising from the random variability of demand and the uncertainty of the lead time. The resulting pattern of stock holding is shown in Figure 11.3 where the solid line represents the stock held, and the broken line the stock on hand (as defined above).

In the two-bin system a fixed quantity is ordered at variable intervals of time; in general to operate such a method needs continual monitoring of all stock transactions and it is only with the advent of the computer that it has become at all widely used.

The re-order quantity q can be evaluated from the expression

$$q = \left(\frac{2DC_0}{C_H}\right)^{1/2}$$

where D is the mean demand for the part per unit time,
          C_0 is the cost of the replenishment order,
          C_H is the cost, per item, of holding the part.

The re-order level M can be calculated from the expression

$$M = DL + k\sigma_D L^{1/2}$$

where L is the mean lead time,
          σ_D is the standard deviation of demand per unit time,
          k is the standard normal variate.

In this evaluation, the cost of stockout is incorporated in the idea of a required *level of service* – an acceptable value of the likelihood that, during any given lead time, demand *will* be met. It can readily be shown[3] that, if the rate of demand for the item can be assumed to be 'Normally' distributed, or approximately so, then the probability that, during a lead-time, demand will *not* be met, i.e. a stockout *will* occur, is a function $F(k)$ of the 'standard normal variate' k, i.e.

$$1 - (\text{Level of service}) = F(k)$$

Thus, if the desired level of service is, say, 99%, then $F(k)$ will be 1% or 0.01, and k is then readily found from the published tabulations of the standardized normal probability density function (see Example 1 at the end of the chapter).

Current stores-control software uses models such as the above to automatically control inventory levels for fast moving spares. In addition, it can monitor changes in demand, and in the other variables involved, and automatically adjust the control levels, i.e. the settings for M and q.

Inventory policies for the control of slow moving spares

It was noted earlier that the greater part of the value, and hence the dominant control problem, of a spares inventory lies in the expensive slow moving parts, where overstocking is not quickly corrected by subsequent consumption. The decision that is then required is whether to hold none, one or – at the very most – two of a given part. Mitchell, working for the National Coal Board of the UK, developed a technique for dealing with this problem[4].

Random failure parts: If demands for a part, although infrequent, occur quite randomly (i.e. they are equally likely to occur at any time) then the probability $P(n)$ of receiving n demands in any given lead time can be assumed to be given by the Poisson distribution, i.e.:

$$P(n) \quad = \quad \frac{m^n \exp(-m)}{n!}$$

where m = mean demands per lead time = $L.D.$

For a re-order level system where only one item is ordered at a time (which would probably be the case with a very high cost item) and demand is *non-captive* (i.e. stockout would be met from another source, e.g. by making – at known extra expense – the spare in the workshop; with *captive* demand stockout would be met by earlier delivery of a spare on order), Mitchell derived the decision chart shown in Figure 11.4. It indicates the value of N, the number of items *on hand* (i.e. in stock plus on order), which will minimize C_N, the average total cost per unit time (of holding and stockout, the cost of ordering being assumed negligible). For points on the line $C_0 = C_1$ equal cost arises if $N = 0$ or 1; along $C_1 = C_2$ equal cost arises if $N = 1$ or 2 (unlike the curve $C_0 = C_1$, the position of this latter curve is a function of L – and is therefore plotted for various values of this).

So, for a given spare, for which L, D, C_H and C_S (the stockout cost) are known, the chart is used as follows. If C_S/C_H and D give a point lying

(a)   below $C_0 = C_1$ then no spare should be held,

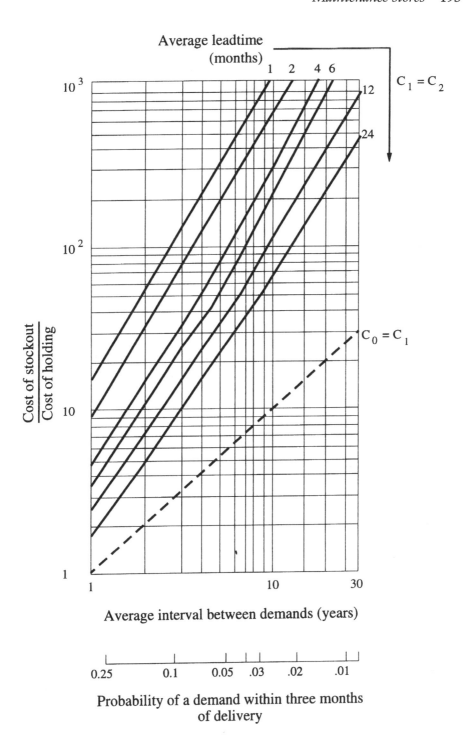

Figure 11.4   Slow moving spares inventory decision chart

(b)   between $C_0 = C_1$ and $C_1 = C_2$ then one spare should be held,

(c)   above $C_1 = C_2$ then two spares should be held.

(See Example 2 at the end of the chapter.)

Wear out failure items. The same chart can be adapted for deciding stock levels for slow moving *wear-out* items. The chart's bottom scale is replaced by one showing the probability that, for a given part, a demand will occur within the first three months following its delivery (given that no demand has occurred while it has been on order). If, for a particular spare, the assessment of this probability, and of C_s/C_H, give a point on the chart:

(a) below $C_0 = C_1$ then ordering is deferred,

(b) above $C_0 = C_1$ then one spare is ordered immediately.

It is recommended that each such decision problem should be re-evaluated approximately every three months (see Example 3 at the end of the chapter).

Mitchell further recommended that, for the purposes of his approach to their control, slow moving spares should be classified as below:

> *Specials* (bought for use on a specified date, e.g. for a plant modification or overhaul). Should be ordered so that delivery occurs as shortly as possible before their use. Clearly, the 'safe' prior interval will be a function of the confidence with which the date for the part's use and the lead time for its ordering are known.

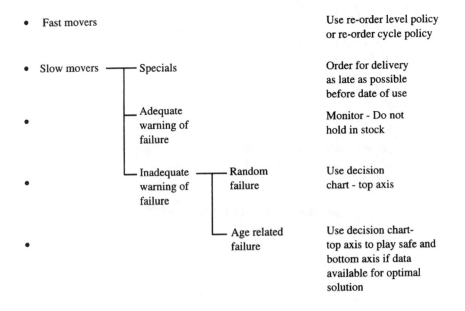

• Fast movers	Use re-order level policy or re-order cycle policy
• Slow movers —— Specials	Order for delivery as late as possible before date of use
• —— Adequate warning of failure	Monitor - Do not hold in stock
• —— Inadequate warning of failure —— Random failure	Use decision chart - top axis
• —— Age related failure	Use decision chart- top axis to play safe and bottom axis if data available for optimal solution

Figure 11.5   Guidelines for spares control policy

Adequate warning items. Condition monitoring (or some other indication of impending failure) can provide adequate notice, relative to the lead time of failure. The part is therefore not held in stock.

Inadequate warning items. For technical or economic reasons no inspection or other technique is available for providing notice of failure. Such parts can be sub-divided, as already discussed, into those that fail randomly (controlled via the first procedure for Figure 11.4 explained above) and those that wear out (controlled via the second).

The author has used this classification as a basis for general spares control guidelines (see Figure 11.5).

System documentation

To manage a stores system of the kind outlined in Figure 11.1 some form of documentation system is essential. Although these are now almost always fully computerized the principles underlying their operation can be most clearly explained by reviewing the various activities of a traditional paper-based system (see Figure 11.6).

All parts must be given a description, a stores code number (see Table 11.1) and a 'bin' location number (see Table 11.2) – 'bin' being used in a general sense for any storage arrangement (shelf, drawer, pigeon hole, box, drum, sack, marked floor space or whatever) that isolates a part type. These are basic essentials for the elimination of duplication, for simpler parts ordering and faster part location.

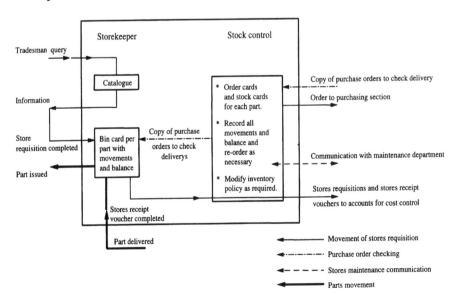

Figure 11.6   A paper-based stores documentation system

Table 11.1 Extract from a simple coding system for spare parts

Two digits to identify major groups, e.g.

01	Abrasives
05	Bearings and accessories
06	Belting and accessories
09	Chain, rope and wire
14-17	Fasteners
19-20	Hose and fittings
34	Packing
36-38	Pump parts
50-53	Piping
56	Valves
60-79	Electrical and electronic
80-99	Plant-specific spares

Two digits to identify sub-groups, e.g.

06-35	Belt - timing
06-40	Belt - transmission
06-45	Belt - variable speed

Three digits to assign unique part numbers, e.g.

06-45-255	Belt - variable speed, Dayeo 46V26

Table 11.2 Extract from spare parts catalogue

Valves, Group number 56	Unit of issue	Cost	Re-order point	Re-order level	Bin number
56-01-150:$\frac{1}{8}$ cock, Air BR 125 SE	each	2.26	2	4	2/04
– 165: $\frac{1}{8}$ cock, Pet BR 125 SE	each	1.75	2	4	2/02

Figure 11.6 models a 'closed' stores system, one in which no part can be delivered or withdrawn without the transaction being recorded. The issue of a part to the tradeforce is covered by a *stores requisition* document (see Figure 11.7 (a)) on which data (plant or unit number, job number, work type, etc.) are also recorded for the maintenance costing system. All deliveries (which should be checked for conformity to specification) are covered by a *stores receipt voucher* (see Figure 11.7(b)).

For each part type the storeman keeps a *bin card* on which he records all transactions (receipts and issues) and the stock balance – many stores also attach a simple *part label* to at least one part as an identification check. All requisitions and receipts pass through stock control on their way to the accounts department and the stock controller records them, and also the various stock levels, on *stock control record cards*. This information is in turn required for the operation of the inventory control policy (see Figure 11.3) and for monitoring changes in demand rates, lead times and so on. Stock replenishment orders are handled by the purchasing department.

ENGINEERING STORES REQUISITION							S.R. No. 0003
Plant description			Plant/Job number				
Location			Cost code number				
Date			Number				
Part number	Description	Quantity		Price		Bin location	
		Req'd	Issued	Unit	Total		
				-			
Tradesman's signature	Foreman's signature	Storekeeper's signature		Stock control entered			

Figure 11.7(a)   Stores requisition

ENGINEERING STORES RECEIPT VOUCHER						S.R.V. No. 0003
Re-order date	Note number		Plant/Job number			
Supplier			Cost code number			
Order number			Number			
Part number	Description	Quantity		Balance	Bin location	
		Ordered	Received			
Received by	Checked by stores	Checked by supervisor		Stock control entered		

Figure 11.7(b) Stores receipt voucher

In most companies the computerized system for stores control is integrated with the systems for general purchasing, invoicing and maintenance activities. The stores system is 'paperless', even down to the use of bar-code readings for registering parts requests and issues. Most stores catalogues facilitate searching for parts under various standard codings and verbal descriptions; parts may also be identified and ordered by tradesmen operating from their own terminals.

Maintenance stores administration

A typical stores organization is outlined in Figures 11.8(a) and (b), the former showing the stores position within the maintenance resource structure (i.e. mainly centralized but several outposts) while the latter indicates the commercial department's responsibility for stores management (i.e. for its budget, for part storage, cataloguing, issuing, ordering and receipt, inventory policy and staffing), the maintenance department itself being responsible for the initial order quantities and for the specification of parts. This tends to be the arrangement because of the advantages of centralization of stockholding, purchasing and invoicing. (Remember that maintenance stores accounts for only one part of the total company stockholding – albeit in most cases the dominant part.) The main problem with this arrangement is the tendency of all maintenance departments to play safe, overspecify and overstock – this is especially so with the slow moving and insurance parts. Periodically, the commercial department may attempt to correct overstocking and in doing so may over-react, leading to eventual stockouts.

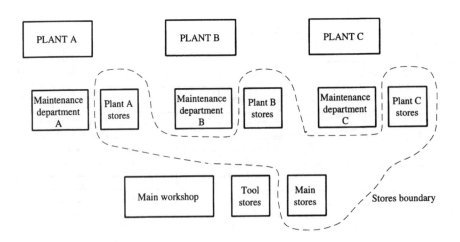

Figure 11.8(a)   Resource structure, showing stores location

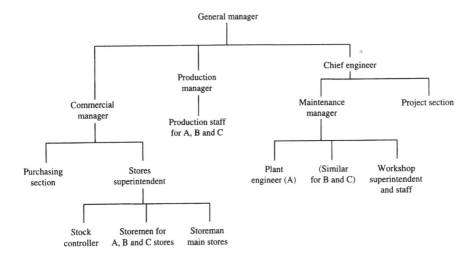

Figure 11.8(b)   Administrative structure, showing centralized responsibility for stores

Where the responsibility for spares management is as shown in Figure 11.8(b) effort must be devoted to:

- clearly identifying the role of the commercial department,
- putting systems in place at the maintenance/stores interface which will ensure that decisions should *always* be referred back to maintenance regarding the inventories of *insurance spares*, and also of *high cost parts* (to ensure they are still wanted),
- ensuring that there are stores staff who are technically competent in the maintenance and engineering areas.

There are many possible variants of the structure of Figure 11.8(b). Some companies, for example, make the maintenance department fully responsible for spares; others adopt the Figure 11.8(b) arrangements but also make maintenance responsible for the inventory policy.

Examples in inventory control

1. Fast moving parts

Calculation of the re-order level M for a particular spare given the following data.

The average demand rate and its standard deviation are $D = 20$ and $S_D = 5$ demands per month, respectively.

The ordering lead time $L = 4$ months and the desired level of service is 99%.

$F(k) = 1 - $ (level of service) $ = 1 - 0.99 = 0.01$.

From tables of the Normal pdf, the standard variate $k = 2.326$, so

$M = (20 \times 4) + (2.326 \times 5 \times 4^{1/2}) = 103$ parts.

2. Slow moving parts – random failure

Estimation of the optimum stockholding for an electric motor, given the following information.

Estimated cost of holding the motor, C_H = £100 per annum.
Estimated cost of not being able to
replace from stock in the event of an
unexpected failure, C_S = £1000.
Average lead time for re-ordering, L = 12 months.
Motor failures occur randomly with an
average incidence D = 0.20 failures per annum.
Thus, average interval between demands $= 1/D = 1/0.20 = 5$ years,
and C_S/C_H = 1000/100 = 10.

These are the co-ordinates of a point on Figure 11.4 that lies above the line $C_0 = C_1$ and below the line $C_1 = C_2$ for $L = 12$ months (NB. The vertical and horizontal scales of the chart are logarithmic).

The decision is therefore to hold one spare motor.

3. Slow moving parts – wear out failure

Estimation of the optimal stockholding for a gearbox given the following information.

Estimated cost of holding the gearbox, C_H = £100 per annum.
Estimated cost of not being able to
replace from stock in the event of
an unexpected failure, C_S = £500.
Average lead time for re-ordering, L = 6 months.

Failure data have been extracted from the maintenance records of a user with many such gearboxes and are prepared as shown in the following table, for plotting on Weibull probability paper.

Time t from new (weeks)	Cumulative percent $F(t)$ of gearboxes failed
120	0.00
160	2.50
200	12.50
240	37.50
280	70.00
320	90.00
360	97.50
400	100.00

The straightest plot is produced (see Figure 11.9) if the guaranteed life t_0 is taken to be about 120 weeks and gives a gearbox mean life of about 260 weeks, or five years (i.e. this is a slow-moving item) and a β-value of about 3.4 (indicating a wear-out mechanism of failure).

For this sort of item the second of the decision chart (Figure 11.4) procedures – involving reference to the scale on the lower of the two horizontal axes – is followed, the position being re-assessed every three months from when the gearbox is new. At each re-assessment it is necessary to re-calculate the probability that there will be a demand for a gearbox within three months of its delivery.

Example: a review at 160 weeks into the gearbox's life.
    If ordered now, delivery would be at $160 + 26 = 186$ weeks.
    The lower scale requires assessment of the probability of a demand occurring within three months (twelve weeks) of that delivery. This is given by the ratio of areas indicated on Figure 11.10, i.e. by

$$\frac{A}{A + B}$$

where the areas A and B represent probabilities. Their values can be derived from the Weibull plot of Figure 11.9, which indicates that the cumulative probabilities at 186 weeks and 198 weeks (i.e. the areas, under the failure probability density distribution, to the left of those points) are, respectively,

$F(186) = 6.5\%$ and $F(198) = 11\%$.

Thus

$$\frac{A}{A + B} = \frac{11 - 6.5}{100 - 6.5} = 0.05$$

Also

$$\frac{C_s}{C_H} = \frac{500}{100} = 5$$

These are the co-ordinates of a point lying just above the line $C_0 = C_1$ on the decision chart. Thus, at 160 weeks into the life of the gearbox the decision would be to buy a spare.

Summary

The most difficult task in this area is managing the slow-moving spares, for the following reasons.

- Such parts typically account for 80% of the cost but only 20 % of the inventory.

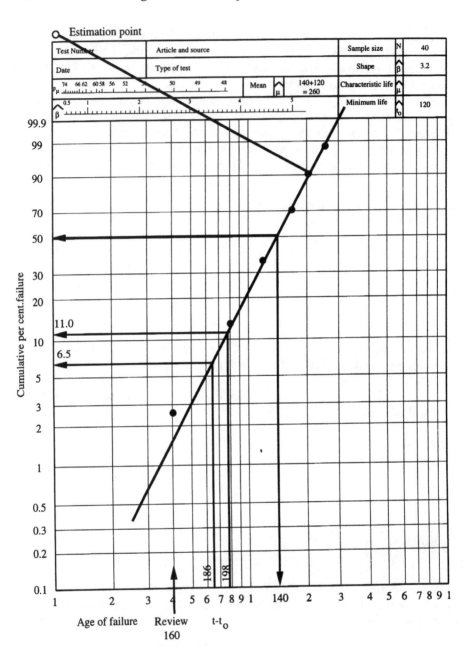

Figure 11.9 Weibull plot of gearbox failure data

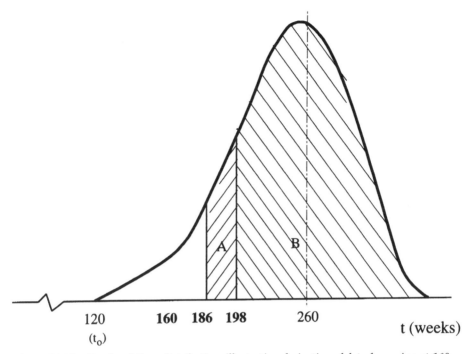

Figure 11.10 Gearbox failure distribution, illustrating derivation of data for review at 160 weeks

- Reliably assessing the likely demand rates and stockout costs is not easy, which limits the effectiveness of the quantitative techniques that can be used for determining inventory policies.
- The division of responsibility for the management of such parts results in 'play safe' initial orders and long-term over-stocking.

A recent audit of a power station stores revealed that slow movers and insurance parts accounted for some 83% of the stockholding costs and that optimization of the stores management policies and practices could achieve savings of £2.25M per annum.

References

1. Kelly A and Harris M J, *The management of industrial maintenance.* Butterworths 1978.
2. Lewis C D, *Scientific inventory control.* Newnes-Butterworths 1971.
3. Harris M J, An introduction to maintenance stores organization. *Terotechnica,* Vol. 1, pp 47-57, 1979.
4. Mitchell G H, *Problems of controlling slow-moving spares.* Opl. Res. Q., Vol 13, 1, p23, 1962.

12
Maintenance documentation systems: What they are and how they work

Introduction

The basic system paradigm that was outlined in Figure 1.1, and briefly discussed in the accompanying text, made clear that some form of documentation system, for recording and conveying information, is an essential operational requirement for all phases of the maintenance management cycle.

Maintenance documentation includes any record, catalogue, manual, drawing or computer file containing information that might be required to facilitate maintenance work. The system is the formal mechanism for collecting, storing, analysing, interrogating and reporting that information and is often constructed so as to facilitate the operation of complex work planning systems of the type that was outlined in Figure 8.6.

Although most current systems are computerized, the basis of their mode of operation has evolved from that of the traditional paperwork system and can be most easily explained by discussing the various components and information flows of the latter.

A functional model

The way in which a maintenance documentation system generally functions is outlined in Figure 12.1, a model which has evolved over a number of years through extensive studies of both paper-based and computerized systems and which therefore illustrates the principal features of both types – features which, inevitably, they have in common.

The system can be considered to be made up of the following interrelated modules:

1. Plant inventory
2. Maintenance information base
3. Maintenance schedule
4. Inspection schedule
5. Short term work planning and control

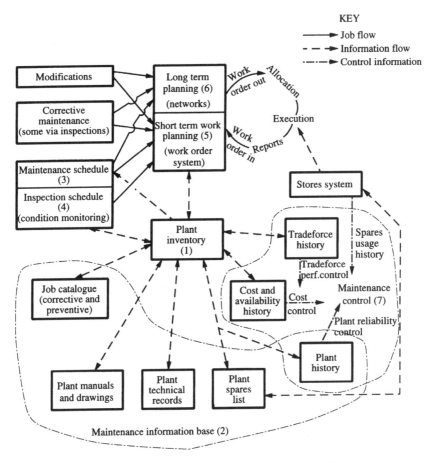

KEY
→ Job flow
- - → Information flow
-·-·→ Control information

Figure 12.1   The maintenance documentation system, a general model

6.   Shutdown work planning and control
7.   Maintenance control

Plant inventory (Module 1)

The inventory can be regarded as 'the centre of the documentation universe'. It is a list of the most basic information about every unit – of machinery, equipment, etc. – in the plant, each unit being uniquely identified by a short description and a numeric or alpha-numeric code which flags up the main way into the system to obtain information on that unit. The list normally differentiates between equipment down to at least unit level and may be complemented by drawings showing the location of every unit listed (see Table 12.1).

Table 12.1 Extract from a plant inventory

Plant inventory number	Unit description	Location drawing	Manufacturer, type and year	Cost
03 - 043	Water discharge system	014C53	Smith Mark IIB 1986	£10,000

A typical coding system could be constructed as shown in Table 12.2. The unit number, 03/F/002, would be used for the on-site identification of the unit; this is essential for work planning, for safety management and for the collection of cost and unavailability data. It should indicate whether the unit is mechanical (numbered, say, in the range 001 to 499), electrical (500 to 699) or instrument (700 to 999), and within each of these divisions the most straightforward numbering is probably in the order in which the units occur along the route of the process. (A separate coding for identifying process flow or electrical flow may already exist, but it will probably not be suitable for maintenance purposes.)

As in the example, an assembly number may be added for accessing data from the information base or for failure recording. This degree of differentiation between equipment is normally only possible, however, if information filing and data handling are computerized. Where large expensive assemblies, e.g. gearboxes, are reconditioned and very possibly re-installed on a different site, or on a different unit, an additional number – separate from the inventory number and uniquely identifying the particular piece of equipment – must be assigned (such assemblies are sometimes called *rotables*). This aids the tracking of such assemblies, the planning of reconditioning and enables a separate history record to be kept. Gearboxes and motors, for example, could be identified as in Table 12.3.

Table 12.2 A coding system

Plant serial number	Plant stream letter	Unit serial number	Assembly serial number
03	F	002	03

Table 12.3 A coding system for rotables

Gearbox	Type	Size	Serial number
G	XX	XX	XXX
Electric Motor	Type	Size	Serial number
EM	XX	XX	XXX

Regarding coding systems in general, Idhammer[1] gives the following advice for ensuring system flexibility:

(a) Keep the spare part code separate from the plant inventory code. The latter should facilitate access to the information base at the point where the coding of the relevant spares is listed.

(b) Keep the drawing code separate from the plant inventory code (but inter-connected as in (a)).

(c) Keep the accountancy code separate from the plant inventory code. All jobs require an accounting coding, but only site jobs need a plant coding for maintenance cost control.

(d) If there are several sites, do not over co-ordinate the coding system.

Information base (Module 2)

For the efficient planning of work it is essential that maintenance-related information is held for each of the units in the inventory, the most important of this being:

- essential technical data;
- spares list;
- drawing records with index;
- maintenance instruction manuals;
- catalogue of standard preventive and corrective jobs;
- unit history.

The information base can be regarded as the sum total of all such data categorized by unit number.

Technical data. An expansion of the basic information in the plant inventory. With a paper-based system, this is usually filed by unit and often sub-divided into electrical, mechanical and instrumentation files. It holds information needed by the planner (e.g. on outside service engineers, on the manufacturer, on the unit specification). Figure 12.2 shows how a 'visible edge' record card can be used for this purpose. The visible edges identify the various units and, taken together, can be regarded as constituting the plant inventory.

Spares, or spare parts, list (SPL). A schedule, for each unit, of all relevant spares held in stores, the stores codings being listed against the unit's plant inventory number (see Table 12.4). Other information, useful in an emergency, could also be included such as the location, on other units, of identical or similar parts. Some companies add to this the spares available from the manufacturer and call the full list the *bill of materials* (BOM).

INSTALLATION		DIMENSIONS		LOCATION			VALUE	
DATE	Sept 1972	WEIGHT	9 Ton	COMPANY	XY Tools Ltd		PURCHASE	£6005-00
ORDER No	664/E	FLOOR AREA	4'.1"x8'.0"	SITE	Spade Forge		DATE	14/8/72
Ac No	-	MAX HT	10'-0"	SECTION	Spade 1	CC14	PRESENT	
DATE MAN'F'D	August 1972	MOUNTING	2'-0"thick Concrete	DATE	Sept /72		DATE	

DESCRIPTION AND REFERENCES	MANUFACTURERS No.	
"Rhodes" Special RF 100 Geared upright open front fabricated press		
Pressure capacity 100 Ton $\frac{1"}{4}$ from bottom of stroke		
90 Ton $\frac{1"}{4}$ " " " "		

MANUAL AND CATALOGUES	
Rhodes RFP 703 (modified)	
	FILE No

DRAWINGS	
Modified Ram Rhodes drawing number 17117	
All other drawings retained by Rhodes	
Prints supplied modified Ram guides JRS 2048	
" " " AYH 289	
Bed plate Print AYH 99	

PLANT DESC.AUX. EQUIPT.
Fixed stroke 4" ram adjustment 4" by hand ratchet lever
Speed 80 S.P.M.
Pneumatic friction clutch and brake. Single stroke and "Inch" cycles foot control
Tecalemit one shot lubrication

PLANT DESCRIPTION	PLANT No. 1650
100 Ton Rhodes C frame press (socket swaging)	

Figure 12.2   A visible edge record card

Table 12.4   **Extract from a spares list**

Plant inventory number	Description	Location	Assembly identification number	
03-043-046	Water pump	Wood handling	P-43-06	
Part name	Quantity	Price (£)	Stores code	Stores location
Pump assembly	2	300	123456	Shelf 1 (main)
Pump housing	1	25	123457	Shelf 2 (main)
Bearing	1	1	123452	Shelf 25 (main)

Table 12.5   Extract from an index of drawings

Plant inventory number	Own drawing number	Drawing description	Manufacturer's drawing number	Date prepared	Date revised
Unit 03-043	0363943	Assembly drawing	2941/1973/350	73.01.06	
	3373735	Pulley drawing	2951/1973/387	73.04.06	

Drawing records and instruction manuals. The drawing index should list the user's and the manufacturer's drawing numbers against the unit's inventory number (see Table 12.5).

Drawings can be held in drop-leaf files, on micro-films, on micro-fiche or – via a document-imaging process – in computer files; maintenance manuals can be held similarly or in a simple library.

Job catalogue. Lists the preventive and standard corrective jobs for each unit of equipment.

The descriptions of the former of these stem from the establishment of the maintenance life plan for each unit, an example of which is shown in Table 12.6. Each of the preventive jobs listed in the example can be written in the form of a job specification (see Figure 12.3) in which each action listed could, if necessary, refer via a code number to the kind of detailed procedure or inspection shown in Figure 12.4. Although not shown, it is now usual for such standard job descriptions to contain estimates of duration and manpower and a list of the spares needed for job completion – such a list being sometimes referred to as the *application parts list* (APL). In addition, these descriptions often indicate the plant status required if they are to be carried out (e.g., major overhaul, weekend shutdown, or on-line) and other jobs that could be carried out at the same time (opportunity scheduling). In the case of major shutdowns the catalogue may include job descriptions linked to bar-charts.

Recurring corrective jobs (sometimes called standard jobs) are specified in a similar way and entered into the catalogue against the relevant units. Because they do not have a frequency of execution their specifications are called up via the unit number when the need for the work arises

Plant history. It can be seen from Figure 12.1 that the history of a unit has a dual function, contributing both to the maintenance information base (e.g. regarding when and how it was last repaired) and also to the plant reliability control system (e.g. facilitating identification of recurrent failures and their causes, see later). The former function assists the planning of work and for this the history is best held in narrative form. Typically, for each job it should include:

- date carried out,
- unit involved,

Table 12.6 Life plan for a crane

	Maintenance life plan	Inventory No: 73-103
Unit description:	Five ton crane	Site: Warehouse (73)
Weekly inspection	Mechanical General Check long travel drive motor for noise vibration and abnormal temperature. Check cross travel drive motor for noise vibration and abnormal temperature. Hoist Check motor for noise, vibration and abnormal temperature. Closely examine rope sheave and hook for damage. Establish correct operation of top and bottom limits. Check security and condition of pendant control.	Electrical NONE
Three monthly inspection	Long travel Check security of motor mountings. Test track for correct operation and check lining wear. Check security of drive shaft bearings. Inspect condition of reduction gears. Cross travel Check security of motor mountings. Test brake for correct adjustment and check lining wear. Hoist unit Test brake underload for correct adjustment and check lining wear. Check gear case for oil leaks. Inspect rope for wear and fraying. General Report on condition of lubrication. Report on general cleanliness of machine.	Long travel Check security and condition of motor leads and earthing. Inspect downshop leads for correct tension and slippers for wear. Cross travel Check security and condition of motor leads and earthing. Inspect catenary assembly for damage and check free operation. Hoist unit Check security of motor leads and earthing. Check condition of wiring to top and bottom limits. Check wiring and push buttons on pendant control. Controller Check condition of all wiring and security of connections. Check for correct and free operation of all relays. Check setting of overloads at 10.8 amps.

Job specification	Plant No.		73 103			
Plant description	Maintenance code		11			
	Job code		Mech/3 monthly (M3)			
Site						
	Week Nos.	8	24	37	1	
Spares required						
Drawings and manual refs.						
Special tools						

LONG TRAVEL

1. Check security of motor mountings.

2. Test brake for correct operation and check lining wear

3. Check security of drive shaft bearings.

4. Inspect condition of reduction gears.

CROSS TRAVEL

5. Check security of motor mountings.

6. Test brake for correct adjustment and check lining wear.
   etc.

Figure 12.3   A job specification

- duration, and resources used,
- condition of unit and details of work performed,
- parts replaced and materials used.

Preventive maintenance schedule (Module 3)

The preventive schedule is formulated from the recommendations of the unit life plans and job specifications (see Figure 12.5). The life plan for the crane shown in Table 12.6, for example, is made up of four preventive jobs of different frequencies (see the three-monthly mechanical service listed in Figure 12.3) and the schedule would programme all such jobs taking into consideration plant and resource availability (see Figure 12.6 in which it can be seen, for example, that the three-monthly preventive work for the crane is scheduled

COUPLINGS, TYPE LB, ASEA, BEK. etc

Inspect for wear on the rubber bushings as
follows. Turn the coupling halves away from
each other. Make a mark straight across the
halves in this position and then turn the
halves in the opposite direction to the
first turn. Measure distance between marks.

Job specification No. 0255

page 1 of 2

Wear

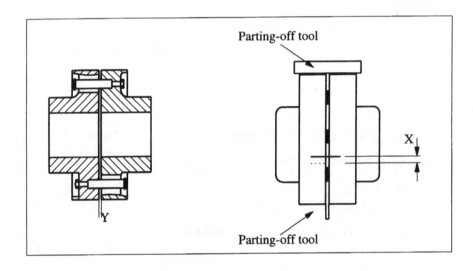

Coupling diameter	Max measurement X
0 - 150	4mm (6mm)
150 - 250	8mm (11mm)
250 - 400	14mm (18mm)

(The values in parentheses relate to
stroboscope measurement)

Measure the distance Y at four points on the
periphery without turning the coupling. Max.
permissible difference 0.1mm for a medium
sized coupling.

Figure 12.4   Detailed procedure

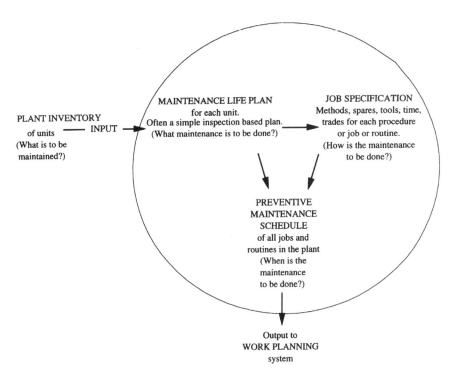

Figure 12.5   Outline of a traditional preventive maintenance documentation system

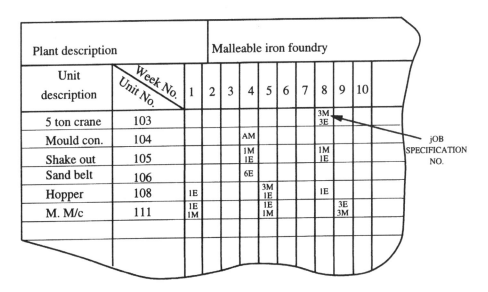

Plant description		Malleable iron foundry										
Unit description	Week No. / Unit No.	1	2	3	4	5	6	7	8	9	10	
5 ton crane	103								3M 3E			
Mould con.	104				AM							
Shake out	105				1M 1E				1M 1E			
Sand belt	106				6E							
Hopper	108	1E				3M 1E			1E			
M. M/c	111	1E 1M				1E 1M				3E 3M		

JOB SPECIFICATION NO.

Figure 12.6   Preventive maintenance schedule for crane

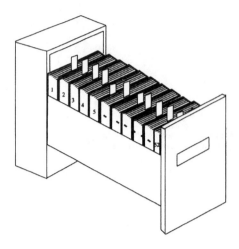

Figure 12.7   Card index

for Week 8, Week 21 and so on).

Where a plant is production-limited the schedule might link several units (and therefore jobs) into a single shutdown, although this could result in a peaky workload and corresponding organizational problems. Conversely, if the plant were sales-limited (or there is an excess, for some other reason, of available windows for maintenance) the scheduling objective might be to smooth the workload.

In most of the traditional, paper-based, documentation systems a schedule as in Figure 12.6 is used mainly for organizing the job specifications into some form of card index (see Figure 12.7). Comprising 52 slots, this can then be used directly for the triggering and control of preventive work, i.e. each week it feeds a tranche of job specifications (each of which is accompanied by a work order on its way to the shop floor) plus a summary, into the work planning system. The index can be updated and re-scheduled as necessary on the return of the job specification cards. Resulting corrective work is noted on the completed work order and this information stays in the work planning system.

In the case of large process plants, the scheduling of preventive work is much more complex and is not always inspection-based. It needs to be integrated over many units of plant and, if a manual system is being used, a chart is used for the initial scheduling and also for controlling the completion of work. At the start of each period the worklist is read from the schedule and the job specifications are then selected from their file and sent, as before, to the shop floor.

Inspection schedule (Module 4)

A life plan for a unit may well include inspections that need to be carried out and hence scheduled, independently of the main maintenance jobs. They can include:

- condition checks and readings carried out by operators,
- inspection routines undertaken by tradesmen,
- condition monitoring routines carried out by technicians using specialist instrumentation.

In their time, the traditional paper-based systems only needed to handle the programming of the less sophisticated of such routines – such as lubrication and simple inspection carried out monthly, say, or more often. The routines, a series of checks, were often listed on a single card which might be sent to the responsible supervisor so that he could organize the tasks into the ongoing workload. Such inspections were subjective and relied for their efficacy on the experience and motivation of the tradeforce.

During the last ten years or so there has been an expansion in the use of condition monitoring techniques such as those based on vibration analysis. This in turn has led to the development of specialist computer packages for scheduling inspections and recording and analysing their results – a process which, from time to time, will trigger off a corrective maintenance job.

Note: The batch chemical plant example discussed in Chapter 9 of Book I could also have been used to illustrate the traditional approach to preventive maintenance documentation. Specifications would need to be written for each of the preventive jobs identified in the life plan (see Table 9.3, Book I) e.g.

- Six-monthly mechanical inspection of reaction unit 03FOO1
- Two-yearly replacement of Type A valves, etc.

An extract from the schedule of these off-line window jobs was shown in Figure 9.10, Book I, and an outline shutdown schedule in Table 9.4., Book I.

Short-term work planning (Module 5)

The principles of work planning, scheduling and control were discussed in Chapter 8, the work planning system being modelled in Figure 8.6. As has been explained, the work order (see Figure 12.8) can be regarded as the vital operational vehicle for this system. The flow of this documentation – the main components of (and aids to) which are listed in Tables 12.7 and 12.8 – is shown in Figures 12.9(a) and (b).

Requests for *emergency* work are made verbally to area supervision who raise work orders. Requests for *deferred corrective* work and for *modifications* are made to the planning office on work request forms. A work order is raised directly or, if appropriate, by reference to the job catalogue. Priorities of such work are decided at a weekly planning meeting and the various jobs entered into the corrective job list.

Preventive work is planned and scheduled as explained in the previous section. Each week, the preventive maintenance system enters a list of jobs into the work planning system where they are considered for the weekly programme alongside the corrective and modification work. Work orders are

Plant description				Work order number			
Plant number		Plant location		Permits			
				P.T.W.	S.F.T.	L.O.A.	S.F.S.
Maintenance cost code		Job specification number		M	L.V.	H.V.	None
				Place of issue			
Coordinating supervisor and extension		Requested by and extension					
				Support and services			
Job/defect description				M	R	C	L	E	HP	W

Stores check		
Special tools		
Transport		
Check list No		
Action by scheduler		
Stores	Work	
Check initiated	programme	
Available	Permits	
Note issued	requested	
Tools requested	Transport	
	arranged	

Priority		Date issued		Date required		Work allocated to		
Action taken (parts replaced)						Job	Time	Date
						Started		
Cause						Finished		
						Multi-trades involved		
Downtime (if any)		Foreman's signature and comments						

Figure 12.8   A typical work order

raised (by reference to the job catalogue) and work that is not to be carried out is re-scheduled using the planning board and job list.

The complete weekly programme for each area or each plant is formulated in terms of resources, plant availability, and the opportunities known to be arising. This, along with the relevant work orders and any other necessary information (such as manual and drawing references, job specification codes, spares requirements) is then sent to the area supervisor, who uses a short-term planning board or a simple allocation board to schedule and allocate the planned and the emergency jobs that come to him direct. Work orders are raised in triplicate, one copy remaining in the planning office, one with the supervisor and one sent as the order to the tradesmen. As the work order is returned through the system the copies can be filed or destroyed. An important point is that for effective cost control the execution of all work should be covered by a work order and a copy of all completed orders should return to the planning office.

Table 12.7 Work planning documentation

Work order	*A written instruction detailing work to be carried out.* The information this might carry is summarized in Table 12.8. When used to its fullest extent it can act as a work request, a planning document, a work allocation document, a history record (if filed) and as a notification of modification work completed. A typical work order (raised in triplicate and in this case not acting as a work request) is shown in Figure 12.8.
Work request	*A document requesting work to be carried out.* It usually carries such information as person requesting, plant number, plant description, work description, defect, priority, date requested.
Job catalogue	A file of job specifications (preventive and corrective) as previously described.
Planning board	For preventive work, a bar chart as already described. For corrective work, a work order loading board covering a horizon of up to twelve weeks, in units of a week and having pockets to allow the work orders to be scheduled into the appropriate week.
Allocation board	A short-term planning board showing men available on each day of one week. Allows jobs to be allocated to men.
Other	Safety permit, stores and tool note, weekly work programme, stores requisition, top-ten report, cost report, history record.

Table 12.8 Information carried on a work order

Planning information
Inventory number, unit description and site
Person requesting job
Job description and time standard
Job specification and code number
Date required and priority
Trades required and co-ordinating foreman
Spares required with stores number and location
Special tools and lifting tackle required   } Usually carried on
Safety procedure number   job specification
Drawing and manual numbers

Control information
Cost code for work type and trade
Downtime
Actual time taken
Cause and consequence of failure
Action taken

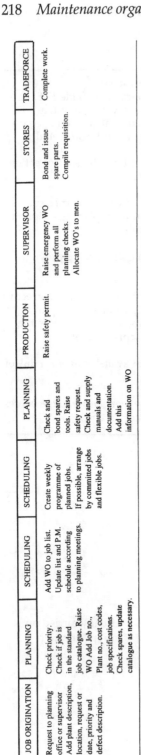
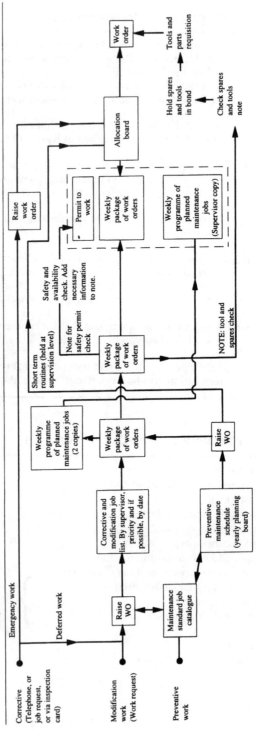

Figure 12.9(a)   Maintenance work planning system

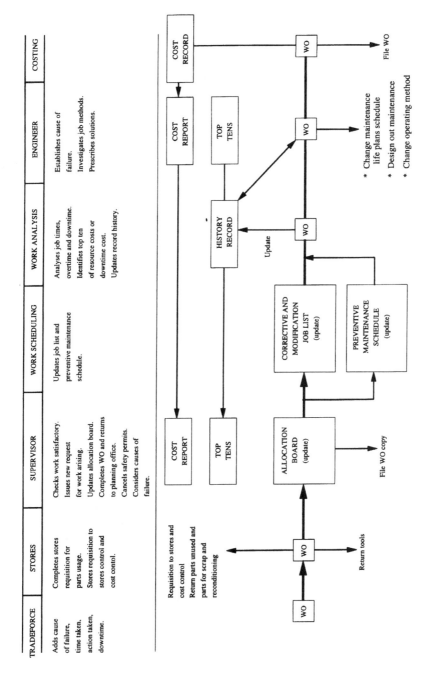

Figure 12.9(b) Maintenance work control system

Long-term (shutdown) work planning (Module 6)

Shutdown planning was discussed in Chapter 9. This deals with the jobs – modification work, preventive work and those corrective tasks that can be held over – that need the plant to be off-line for an extended period. Thus, if the shutdown was to be undertaken under a paper-based documentation regime a simple file to identify and note all such jobs would be needed, which could be categorized according to the units they were to be carried out on.

Shutdown work tends to be different from ongoing maintenance in that the overhaul of even a small unit might consist of hundreds of interrelated jobs. The wall mounted bar-chart is the usual vehicle for the scheduling of a small overhaul. The jobs on the chart can be specified, linked via a bar-chart code, and noted in a shutdown job catalogue. In a large plant where there are many such interrelated overhauls the scheduling would need to be undertaken via a computerized critical path analysis. The execution and control of the shutdown, however, should still exploit the work order system so that day-to-day cost control and history recording may be sustained.

Maintenance control (Module 7)

Much information transfer, storage and analysis is needed to facilitate the various maintenance control systems that have been outlined. It will be instructive to identify the information that is critical for each of the main control systems and also to look at some of the control reporting documents that have been used in paper-based systems.

Work control. The completed work order provides information on the status of each job (completed, delayed, etc.), the resources used and actual time taken, allowing the programme, job list and preventive schedule to be updated. Together with information on tradeforce availability levels (derived from data on sickness and leave) this facilitates control of the work flow (see Figure 8.5). Reports identifying the man-weeks of work in system, categorized by priority and trade, can then be drawn up. Reporting on tradeforce performance is facilitated if work-measurement-derived standard times are estimated for each job going into the system.

Plant reliability control. The key document here is the *unit history*. As well as providing information that guides the execution of future jobs it also has a maintenance control function (see Figure 10.3), i.e. it facilitates the identification of problem units and the diagnosis of the causes of failures. The history record for each unit should therefore contain such information as:

- failure date and/or hours operated to failure,
- duration of failure,
- item/component affected,
- probable cause of failure (see also Figure 10.2),
- production throughput lost.

Description						Plant No.
Date	Item		Defect	Cause	Repair action	Time taken
8.11.79	C4152	SOLENOID REPAIRED (7907)			0.5	0.5
24.11.79	C6356	SCRAP REMOVER REPAIRED (7907)			18.0	18.0
26.11.79	C4169	SOLENOID REPLACED (7907)			1.0	1.0
12.1.80	C6611	SCRAP REMOVER REPAIRED (7907)			17.0	17.0
14.1.80	C4199	FAULTY SOLENOID (7907)			2.0	2.0
26.1.80	C7077	FILTERS CLEANED ON AIRLINE (7910)			4.0	4.0
8.4.80	C5172	SCRAP REMOVER REPAIRED (7907)			6.0	6.0
14.4.80	C5191	SOLENOID REPLACED (7922)			1.0	1.0
3.6.80	C4140	REPAIRS (7909)			3.0	3.0
28.6.80	C9625	NEW GAUGES M/F (7923)			3.0	3.0
22.7.80	C5066	SOLENOID CHANGED (7907)			8.0	8.0
28.7.80	C5072	SOLENOID CHANGED (7907)			1.5	1.5
20.10.80	C5210	SCRAP REMOVER REPAIRED (7907)			1.0	1.0
13.4.81	C7814	NEW SOLENOID FITTED (7907)			1.0	1.0
23.5.81	D1583	SCRAP REMOVER REPAIRED (7920)			12.0	12.0
11.6.81	D2210	MADE & FITTED NEW CYLINDER ROD, REPLACED WORN PARTS WITH PARTS FROM SPARE SCRAP REMOVER (7907)			4.5	4.5
27.6.81	D2106	SOLENOID VALVE CHANGED (7920)			1.0	1.0
17.8.81	D2392	JAWS ON SCRAP REMOVER NOT CLOSING - SHUTTLE VALVE BLOCKED - DIRTY AIR LINE CLEANED OUT			1.0	1.0
17.8.81	3MINSP	NO DEFECTS				
22.8.81	D1001	SCRAP REMOVER OVERHAULED (7926)			32.0	32.0
	Mechanical history record	RHODES PRESS			1650	

Figure 12.10   Extract from a history record

This can be provided via completed work orders, shift reports, defect reports or downtime records. Even with information as limited as this the reliability control system could generate lists of items ranked according to, say, mean time to failure, mean time to repair, or repair hours. For the more troublesome items thus identified, information on item defects and probable causes could then be interrogated to assist the prescription of corrective action. The simplest form of history record is shown in Figure 12.10 and would operate with the type of system that was modelled in Figure 12.9(b).

Maintenance cost and availability control. The function and operation of this system was explained in Chapter 10. Put simply, it depends for its effectiveness on a true history being accumulated of maintenance costs and plant availabilities (and other parameters of maintenance output) for each unit of plant. The main documents used to collect this information are:

- work orders or time cards (which provide data on the man-hours spent on each unit):
- stores requisitions and material purchase documents (for data on the parts and materials used on each unit),
- downtime record cards (for data on downtime, availability, and output for each unit).

The information can be held in a costing record *for each unit* and can facilitate the production of reports on, for example, total maintenance cost and achieved availability per production period per unit. Such data can then be built up to enable figures on total maintenance cost to be set against figures on total output, *per production line* or *per plant*.

Additional reports can be generated by dividing the total costs into preventive, corrective, mechanical, electrical, instrumentation, manpower, material and so on, figures which can be compared with target values or ranked to highlight problem areas.

Concluding observations

The main aim of this chapter has been to develop a general functional model of the maintenance documentation system, to enable the reader to better understand his own documentation system or the computerized ones that are available and which will be discussed in more detail in the next two chapters.

The functional model has been used by the author in the following ways:

- as part of his technique for auditing a company's maintenance documentation system;
- as the basic model upon which to structure a company's *user specification* if it wishes to update its computerized documentation system;
- to guide the construction of a questionnaire for evaluating maintenance documentation software.

Very few companies now use paper-based systems. They required considerable clerical effort for those functions involving data storage, retrieval and analysis, e.g. the provision of an information base (Module 2). Where such systems were used the most effective part was the preventive maintenance schedule (Module 3).

Reference

1. Idhammer, C., *Maintenance course notes for developing countries*. M Gruppen, Fack 1213, Lidingo, Sweden (c. 1980).

13
Computers in maintenance management (i) – an introduction

Some recent history (and a little futurology)

The image of computing power projected by films made during the 1970s and 1980s was of vast air-conditioned rooms full of whirring tape-drives stretching into the distance, and tended by armies of technicians in white coats. Such installations cost millions of dollars, and their running costs were astronomical. Data was held on large reels of magnetic tape, and reports were often run overnight because of processing and performance restrictions. Software was usually specific to the machine manufacturer, and was maintained in-house by large teams of programmers. Computing was very much a rich company's prerogative, and those who wanted to take advantage of it were faced with daunting entry costs.

The advent, in the late 1970s and early 1980s, of the microcomputer (what we now refer to as the PC) struck a blow which the computer industry was slow to feel, but which eventually revolutionized both that industry and business in general. The first of these machines were considered by many to be toys with little commercial use (hence the word 'micro') and no future, a view held by some of the largest computer manufacturers of the day, to the extent that many of them left the market to young entrepreneurs who established companies like Commodore, Apple, Microsoft etc. By the time the big players realized their mistake, some of these small ones had grown to a significant size and some, like Microsoft, would come to dominate an entire market sector.

During the 1970s and early 1980s, the main processor for a typical maintenance computer installation would fill a large room that would need to be hyper-clean and air-conditioned. Access to the system would have been via dumb terminals; documentation, drawings, and so forth would be referenced to hard copy files; and the whole thing would be looked after by a specialist team of full-time technical staff.

Compared with modern computing installations, these systems were slow to use, could only be operated by dedicated experts, and were inflexible. Functionally, these early installations were little more than vast data-storage devices with the ability to schedule – and then print – job cards, reports etc. Machines from this era

---

* Contributed by Peter Bulger, Associate, IMMS

are now more frequently seen in museums.

In complete contrast, today's maintenance computer installation (a DEC Alpha cluster, for example) will probably take up no more space than the average office desk (in fact, will probably look like one), will be roughly a hundred times more powerful than the original at a fraction of the cost, and as each year passes will be rendered increasingly obsolescent by yet more powerful upgrades.

During the last five years or so a further revolution in this technology – one of almost equal significance – has put orders of magnitude more processing and storage power within the reach of almost everyone; children are growing up with a technology which would have been inconceivable ten, or even five, years ago. Maintenance managers can now contemplate relatively modest investments in hardware and software to achieve levels of sophistication never previously thought possible. However, this has raised much more significant organizational and cultural problems (which will be dealt with in Chapter 14).

What then is currently available, and how much of it should the maintenance student or practitioner be expected to understand? It is fair to say that ignorance of technology is unacceptable in the modern business world. Indeed, it is often said with some conviction that a desk-top computer is an essential business tool. The requirement now is not simply the mechanization of paper systems, but the ability to make money out of the information contained within the company database.

The guts of a modern computer are now so small that a simple box to contain them would be too small to provide a keyboard large enough for human fingers to operate. Computers have progressed from dependence on the thermionic radio valve of the 1940s, to exploitation of the first transistors, and now to progressive micro-miniaturization using silicon chips which will soon hold tens of millions of transistors on a single fingernail-sized module.

It is now possible for practitioners to concentrate entirely on the functional capability of any system, without having to worry too much about technical know-how or programming skills.

Modern PCs are extremely reliable and have access to a huge range of ready-made business software. A modern maintenance manager might reasonably be expected to operate a maintenance management system, use spreadsheets, produce customized print-outs and management reports containing graphs as well as text, provide detailed financial reports, and be able to monitor the availability of the plant under his control, the performance of the workforce and the state of his budget. In addition, he should expect to integrate stores, procurement, and general ledger facilities, as well as incorporate electronic drawings and still and video images, together with electronic mailing and access (via Internet and other such providers) to specialized databases throughout the world. All of this from his desk, his home, or his car!

By the time this book is published the current (and fairly new) 586 or

Pentium chip will have been superseded by the P6 processor. The Gartner Group have predicted that by 1999 PCs will support video conferencing and voice technology for a working population which will be 36% 'mobile'.

This paints a picture of a maintenance practitioner rather different from the traditional image. He must be able to understand the business benefits of technology without being seduced by the more colourful but less useful facilities – he must be able to distinguish between software and vapourware! What is useful from what is pretty, and what the hard cash benefits are from any investment in computing. He should also be able to exploit the technology to re-engineer basic business processes, by drawing on simple techniques (see the later section on networking).

The basics of computer hardware

The anatomy of the PC. Comprising several discrete parts (see Figure 13.1) its name will often give an indication of its properties. Let us start with the 'engine' that drives every machine.

The processor. Over the last few years processor development has moved very quickly through a number of significant development phases which have been marked by the release of successively more powerful generations, starting with the 286, and graduating to 386, 486, 586 or Pentium (P5), and now P6 (codings that are often found on the nameplate of the PC). However, the

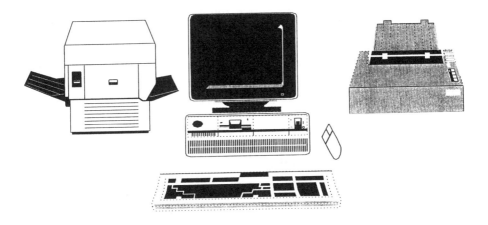

Figure 13.1   The basic parts of a PC system

performance of the PC will depend on two things – firstly the power of the processor (as indicated by the above codings), and secondly the speed at which information moves through the processor, which is expressed in megahertz (MHz) and typically will have values of 33, 66, 100, 133, etc. Generally speaking, the higher speeds are only required for systems which carry out complex scientific calculation, real-time engineering monitoring, or manipulation of graphic or video images. The lower speeds are perfectly adequate for word processing, spreadsheets, and running most commercial maintenance packages.

The memory. This is often referred to as *RAM (random access memory)* and is measured in megabytes (Mb). The bits and bytes terminology which is often scattered throughout the literature and other 'computerspeak' documents refers to the binary codes used within the computer for all its operations. A bit is shorthand for a binary digit, which can be either zero or one. A byte is usually eight bits, which is roughly equivalent to a character on the screen, e.g. the number 52 would be represented by the binary code 00101100.

The memory is a temporary holding area for whatever you are working on at the time. If you are producing a document, or entering data, it is held in memory. If the machine is switched off, whatever is in memory is lost. Many software packages will require a certain minimum amount of memory to be made available for functions which need to be held there. The bare minimum at the time of writing would be 4 Mb, but as soon as more than one application is being run, within 'Windows' say (see the later section on software), then the minimum required will likely be 8 Mb, or 12 Mb for machines that are manipulating very large spreadsheets, etc. Memory can be added to the computer fairly easily by inserting more memory modules and most PCs allow for such expansion.

The main information store. PCs would clearly be of little use if all the information disappeared when the machine was switched off. Long-term storage of data is undertaken on magnetic discs which are either removable (when they are termed *floppy*, although they aren't any more!) or fixed (when they are called *hard*). Removable discs are useful for passing information around, but tend to have a much shorter life than fixed ones. The two can be used together very successfully if the floppies are used to store back-up copies of what is held on the fixed discs. Currently, floppies will hold up to about 1.5 Mb of data whereas fixed discs will easily hold as much as a 1000 Mb. The capacity of hard drives has increased steadily over the years and is likely to increase further.

These discs spin constantly within the PC when it is switched on, with data being read and written by means of an arm which moves across the surface. As this is a mechanical process it is usually the slowest operation in the system. Most new machines also come equipped with a CD-ROM (compact-disc read-only memory). These look just like the music CDs that everyone is familiar with, the advantage being that they can store vast amounts of information – e.g. all the manuals for a large system, or the entire contents of the Encyclopaedia Britannica – and are relatively indestructible. As the name suggests, they are only able to be read, and not written to, although this is changing; in future they may well be used for data storage and thus supersede the magnetic systems for doing this. Within maintenance management, CD-ROMs could be used very effectively for the mass storage of plant records, drawings, suppliers parts list, training and operation manuals, and any other material containing images and text.

Everything discussed so far is contained within the PC 'box' that sits on or under the desk. Getting information in and out requires a keyboard, a screen, and a mouse. Despite many attempts at developing alternatives, the computer keyboard is still based upon the old manual typewriter QWERTY layout. Systems differ as regards the layout of the other, number and function, keys. In specific packages these are often used to initiate functions like 'saving', printing, etc., but with the introduction, into Windows, of *icons* (little pictures representing anything from a simple function like 'Print' to an entire software package) these function keys are now less widely used. The other input device that is now in common use is the *mouse*. Used to move a pointer around the monitor screen, it is a simple 'point and click' method of selecting functions, changing fonts, altering margins, opening and closing files, and moving and editing text. The mouse can be used for everything except inputting text and numbers. On some portable computers, the mouse is replaced by an in-built trackerball which is rotated to move the pointer on the screen.

The computer screen (or visual display unit, VDU). There are very few systems which do not utilize colour, and increasingly the demand is for a 'graphics quality' display (which, currently, would be a *Super VGA*, or *video graphics array*). The quality of the picture on the screen is determined by two factors: screen resolution and the PC's internal graphics capabilities. Screen resolution is measured in terms of the dots (or *pixels*) that make up the display; the more dots, the better the resolution. Currently the standard is specified as 1024×768 but this will change with advancing technology, which is also likely to make the display unit less bulky. To get the best out of a particular specification of screen, the PC needs to contain the appropriate graphics capability. Screen sizes vary from about fourteen inches upwards but this dimension is usually measured from corner to corner as with TV screens, which can be somewhat misleading. It pays to put some effort into choosing the right display for your use, particularly if you are going to spend a lot of time in front of it every day. For specialist applications there is a range of touch sensitive screens, where commands are inputted by touching boxes or pictures displayed on them. This is particularly useful where the work environment is not clean, or in areas exposed to the weather. It can also be used to speed up the 'navigation' around a particular system, where shortage of time precludes the operator from continually looking away from the screen in order to use the keyboard.

Printers. Change seems to occur as rapidly in this technological area as in any other . The days when printers were like typewriters have long gone. High quality colour – or black and white – printing can now be obtained at relatively low cost using either the *laser* or the *ink-jet* techniques. The choice of printer is really determined by the volume of paper you expect to put through it. Generally, laser printers are best suited for applications where hundreds or even thousands of copies a day are required, while ink-jet or ink-cartridge printers are fine for lower volumes. Most printers now operate using plain photocopier paper.

Networking computers

One of the most useful developments in personal computing has been the ability to share information and communicate electronically with other computer users in a 'network', of which there are two main types, known as *peer-to-peer* and *server* networks, respectively. The first of these allows each machine on the network to have equal status, with each performing a similar task. Such networks are typically very small (about ten users) and are not efficient at either sharing software or files of data.

For serious business or commercial use, like maintenance management in a building, factory or power station, a server-based network will be required (see Figure 13.2). These can typically connect many hundreds, or even thousands, of users together and are controlled by one or more central machines (known as file servers) which will contain programs which are required to be shared, and will hold all of the files of data for every user on the network. The server will undertake many system management functions, like controlling security of access, keeping users' files separate so that only authorized access is allowed, sharing access to common software, and routinely backing up data storage to protect the system in the event of a disc, processor, or power failure. A maintenance management system will certainly need to run on such a network, with many users having access to various parts of the system under the control of the server.

There are several variations on the theme of computer networks, and in recent years much interest has been centred around the so-called *client–server* networks. The Gartner group defines these as the 'splitting of an application into tasks that are performed on separate computers, one of which is a programmable work-station, e.g. a Personal Computer'. Figure 13.3 shows some of the variations that are possible within a client–server environment.

In general, a large network may comprise a number of computer applications (types of program), which may be split up between a number of file servers, each one specializing in a particular function, but all

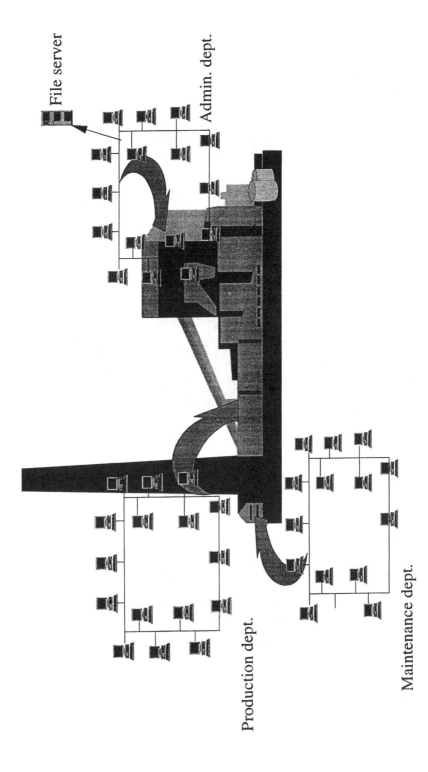

Figure 13.2 Local area network at three locations on a site, controlled by a single file server

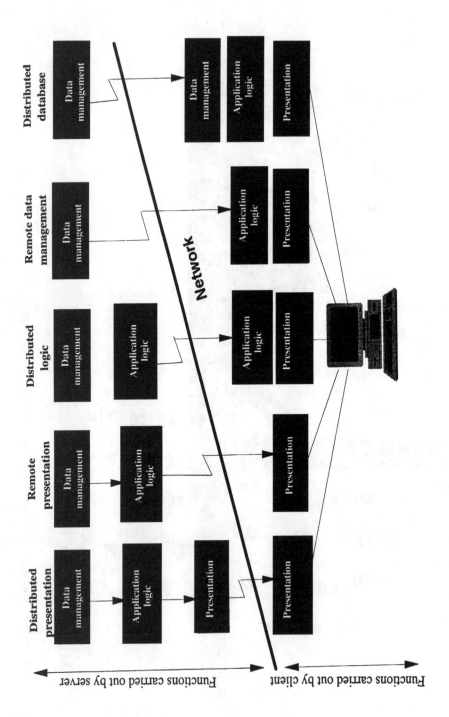

Figure 13.3   Varieties of client–server environment

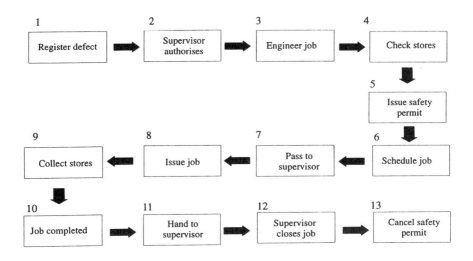

Figure 13.4   Life cycle of a defect card

of them interconnected. This all sounds very complex, and indeed it can be, but the technology is well developed, and the solution suited to any particular application will depend on a number of factors. The reader should bear in mind that the solution is needed to help with the job, and should avoid getting too seduced by the technology; salesmen will always try to maximize their sales, that is their job.

Networking is what really bring systems alive, and it will often offer the ability to send and receive electronic mail and faxes from the desktop. Networks are particularly useful for circulating material for comments, approvals, etc. A large engineering drawing can be circulated using an electronic distribution list, picking up comments (voice or text) and alterations prior to being issued in approved form, and all without a single paper drawing being produced. Such routing functions are often used to proceduralize processes within the business, and can be used to increase efficiency.

Figure 13.4 is a simple flow diagram of the life cycle of a defect card operating within a manual system. Steps 1–4 represent physical movement of the job card, followed by a telephone call to stores. Four more physical card movements take place before the card is handed to the tradesman. Once the job is complete, two more moves occur before the job is completely closed. These moves, of course, do not take place all at once. In all probability this job card would lie in in-trays or couriers' baskets for perhaps a day or even longer.

Contrast this with an electronic work flow based on a modern maintenance management system (see Figure 13.5). Not only are there fewer actual steps in the process, but all movement takes place from one electronic account to another. The engineer responsible for the work simply looks

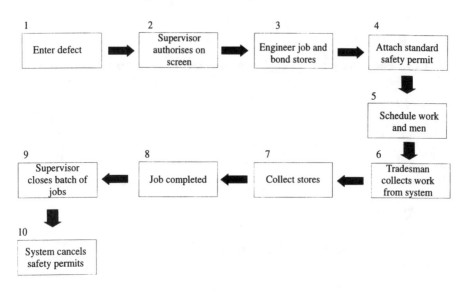

Figure 13.5   Computerized work-control

up all the defects waiting for his approval, makes electronic enquiries of the records of past work held in the system, picks out the work routine that is closest, modifies the instruction, checks the stores module for the relevant spares availability and books those spares for the job. The same process is followed for the safety assessment, in which standard isolations or safety permits are retrieved from the historical records. The tradesman logs on to his account and gets a list of work to be done, together with job instructions and safety documents. By the time he gets to the store, all the material is waiting, chosen by the storekeeper from an electronic picking list issued electronically once the job has been safety-approved.

Wide area networks allow people to be connected together in different parts of the country, continent or world, using a variety of tele-communication methods from ordinary telephone lines, to fibre-optic cable and satellite links (see Figure 13.6). The Internet brings the ability to pick up information which is in the public domain anywhere in the world for incorporation into a company's own documents or programs. Programs made available in this way are often referred to as *shareware*.

Mobile computers

All of the above applies equally well to portable computers. These machines are being progressively reduced in size, a process which is limited only by the size of the batteries required to power them, and of the keyboard – which itself is governed by the size of the human hand. To overcome some of these limitations one manufacturer has designed a

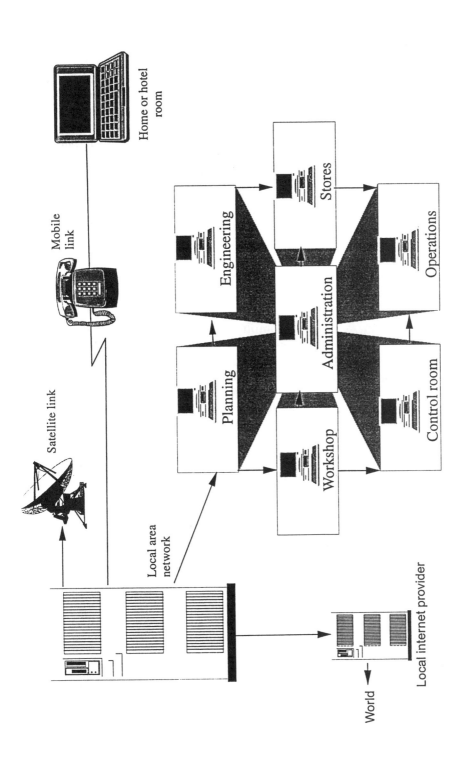

Figure 13.6 Wide area networking

portable machine with a keyboard which folds out for use. In the future, direct voice communication may well provide the ultimate solution, although the limitation then will lie in the size of display which can be read.

The convergence of computing and digital telephony will ensure that the full capability of the maintenance management system will be at hand, no matter how mobile the user, and this will be limited only by the system's ability to lock into the signal from a mobile phone! However, for the most part, the type of systems required within most maintenance functions will continue to use networked static machines distributed around the plant, unless, of course, the plant itself is mobile! In a recent installation concerned with the maintenance of warships, satellite technology was utilized to identify potential fault conditions, maximizing the amount of work which could be undertaken at sea. It is always worth remembering that the application should always dictate the technology, not the other way around.

The basics of software

When computers were first developed, anyone who wanted to make the computer do even the simplest of tasks had to write a detailed list of instructions in a code which the computer could 'understand'. Because computers operate via millions of binary instructions the original programs consisted of hundreds or thousands of lines of these instructions, which of course looked nothing like English (or any other language). Clearly, progress in using computers would depend as much on developments in software as in hardware. Software has developed along four broad lines; in operating systems, in applications, in computer languages, and in databases (see Figure 13.7).

Operating systems. Software which controls the operation of the computer, one of the earliest such systems being known as *DOS* or *disk operating system.* Such software became essential once the first mass-storage devices (disks) were invented. These devices worked rather like an old fashioned gramophone record – with tracks which contained recorded information, and a reading head (like the needle of the record player) which moved from track to track. DOS contained ready-made routines for moving the 'needle' around the disk, picking up or laying down data. In addition it contained ways of transferring information to the screen, or picking up input instructions from the keyboard. This began to make programming much easier, allowing the programmer to concentrate on making the computer produce the desired result without having to spend enormous effort on just making the various bits of machinery connect together. As computers began to use internal memory, routines were added to DOS to handle the allocation of this – allowing work to be done without being continually overwritten.

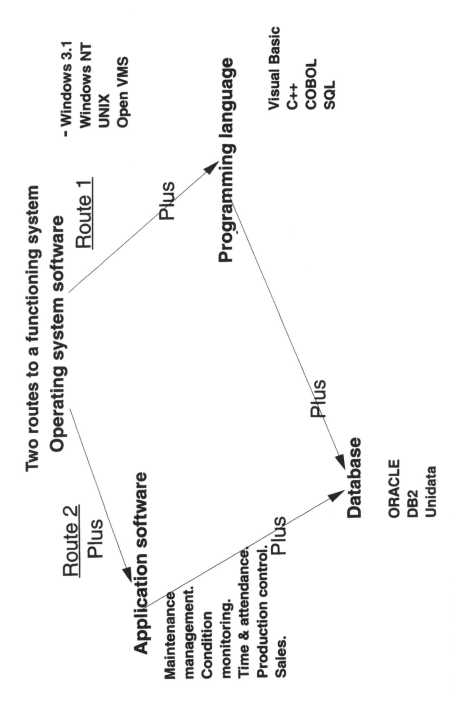

Figure 13.7 Basics of computer software

Operating systems have become much more advanced since DOS was created – driven, mostly, by advances in technology. When DOS was originally written, the maximum amount of memory it could handle was 64 kb, which at the time seemed a great deal. As memory capacity increased beyond this level, DOS became unwieldy.

Perhaps the biggest change came about with the development of *Windows* and what became known as *GUI* or the *graphical user interface*. Screens that were full of characters, words, numbers, etc., were replaced by screens bearing pictures or icons. In DOS a print command consisted of the word 'print' typed into a line of instructions in the program or on the screen. In Windows the command 'print' is simply a tiny picture – contained within a small box – of a printer. When the cursor or pointer on the screen is placed within the box (by moving the desk-top mouse), and a key pressed or the mouse button 'clicked', the box looks like a button that has been pressed, and the command is executed.

The introduction of Windows-based computing enabled the designers of software programs to use the same tool kit of parts, icons, functions, etc., provided by Windows. This meant that different applications running under Windows had a similar look and feel, a degree of commonality. Before the development of the Windows operating system users could only operate one software package at a time. So, if a word processor was being used, and the user decided he wanted access to data in another package, say a spreadsheet, then he would have had to exit the first package, load the second, get the data, exit the second and re-load the first package. Windows allowed several applications to be run at the same time, each in its own 'window', the user being able to switch between them at will.

There are many operating systems now available, including Windows 95, Windows NT, IBM's OS/2, DEC's Open VMS, and UNIX. The UNIX system is very widely used, partly because programs and packaged software written for this operating system are very *portable*, i.e. they can be run on almost any type of hardware It is especially useful in large multi-user environments, where the system is being asked to do a number of simultaneous tasks (which is known as *multi-tasking*).

Application software. In the early years of computing, if a company wanted to computerize its stores system, it would have to write all of the programs necessary to carry out the functions. They would have to be specified in the greatest detail so that programmers could convert them into computer code. The functioning of the programs would be 'hard coded'. In other words, if a user decided he wanted the system to operate in a different way, however slight the difference, extensive rewriting would be required. Considerable testing was then needed to eliminate faults in the programming, or *bugs* as they became known. Any subsequent changes meant a new round of *de-bugging*. Altogether, it was an extremely expensive and time consuming process, and prevented large businesses from changing quickly if their operating conditions demanded it, and all but excluded small businesses altogether.

It soon became clear that most business or commercial functions – like stores control, accounting, payroll handling, etc. – contained a large body of common operations regardless of the industry in which those functions were carried out. Standard *packages* of software began to appear on the market, and gradually overtook the 'bespoke' software. Some of the earliest packages were designed to allow the computer to be used as a text writer. These became known as *Word Processors* and introduced huge savings in big companies where all letters, reports, etc., were produced in large typing pools. The typists who had been employed for this became some of the earliest victims of computerized automation.

The current market is now almost completely dominated by packaged software. There are a number of reference books which contain lists of packages for every conceivable industrial application, these being known by the generic title of *application software*.

Programming languages. Computers, like people, can communicate the same or similar concepts or actions in a variety of different languages. It is said that some spoken languages are more precise than others, Latin for example. Also we sometimes have to invent special languages, mathematical ones for instance, to communicate certain kinds of science. With computers, programming languages have been developed which allow simple commands to represent many binary operations. So when a programmer writes the command *'get x'*, the computer translates this into a string of binary notation which will involve obtaining a value from somewhere within its store, and preparing to carry out some further function.

Many hundreds of computer languages have been invented over the years, some more suited to commercial programming, e.g. COBOL, others to scientific calculation, e.g. FORTRAN. These have been overtaken by languages which begin to look much more like spoken ones, i.e. like English. A maintenance practitioner may well spend an entire working life using computers, but may never need to know how to program them.

Databases. So far, the way that data is structured or stored has not been discussed other than to say to say that it resides on a hard disc. If it were recorded in a random fashion without a logical organization we would never be able to retrieve it. In the early days of computing, data was stored rather like a paper record, starting at the beginning of the file, and proceeding through to the end. To find something, the file was read sequentially until the data required was detected.

Modern systems tend to use what is called a *relational database* (another name for a data file). The data is organized in three dimensions with each data item linked or indexed to other related data items. This means that with a single reference, all data which is related together can be found much more quickly, because it is all linked. There are many such databases, one of the best known being ORACLE. Again, as with UNIX, ORACLE is very portable and will run unchanged on a large number of different hardware platforms.

This topic is a very large one, and for that reason it has been possible to communicate – in a single chapter – only the barest of outlines. Nevertheless, it is hoped that the reader who is contemplating obtaining and/or using computerized maintenance management systems, and who is being exposed to the subject for the first time, will have acquired some helpful basic insights.

14

Computers in maintenance management (ii) – their selection, implementation and use

Introduction

It is no longer the case that companies have to ask the question 'Why computerize?' The complexity of business is now such that it is in data – or more precisely, in information – that competitive business advantage is to be found. The mere ownership of the latest technology is only the entry level qualification. In a recent (1996) London Business School survey of the state of European manufacturing, automation and information systems consistently appeared in the top three priorities for achieving business vision (see Table 14.1).

The best-performing companies, often described as 'world class' or 'best in class', make effective use of computing to support high quality manufacturing processes. This means that an engineer who wishes to become a maintenance specialist must be both computer literate and aware of the functional management opportunities which systems provide. The role of the maintenance engineer in today's industrial setting has changed considerably from what it was some twenty or so years ago – not only because of the advent of computer technology, but also because of the introduction of new production techniques – many of them Japanese in origin – such as JIT, TPM, TQM, Kaizen, etc., whose main focus is reducing waste, improving efficiency, and increasing profits.

Perhaps the most significant word in the title of this chapter is 'management', for this is the whole focus of computerization within the

Table 14.1 Ranking of factors in productivity growth and investment

Indicators of investment	GERMANY	HOLLAND	UK
Automation	1	2	3
Information systems	1	2	3
Indicators of productivity			
Productivity growth	1	2	3
Low product cost	3	2	1

*Contributed by Peter Bulger, Associate, IMMS

maintenance function. Indeed, the purpose served by computerization is now more often referred to as *asset management*, because of the need to aggregate historical costs, and use them in the complex financial decisions associated with replacement or refurbishment of major plant. From the previous chapter some insight will have been gained into the way in which computing power and cost has changed over the last few years. So much so, that the power of a mainframe system, that cost hundreds of thousands of pounds a few years ago, is now available in a box which sits quite comfortably under the desk and is costed in tens of thousands, or even thousands, of pounds. It is said that the exercise of power brings with it responsibility and nowhere is this more true than in computing. The powerful imagery available on modern PCs can often seduce the unwary user into thinking that 'the medium is the message'. If common pitfalls are to be avoided, it is as well to develop a healthy scepticism when dealing with this technology.

The need for change

Although some of the important issues involved in choosing, implementing, and using these new technologies will be spelt out in this chapter, reference will also be made to the wider implications of plant or business efficiency. The maintenance practitioner is likely to be involved throughout his career in re-engineering business processes, applying some of the techniques already mentioned and using computer technology as the facilitator. Computer systems now play a central role in managing all businesses and nowhere is this more true than in the production – and therefore the maintenance – environment. The traditional internal wrangles between maintenance and production must now be seen as anti-competitive behaviour and will stand in the way of a company developing a world class culture.

Today's maintenance professional must take a more strategic view. It is no good having the best product in the world if the plant which produces it is not reliable enough to meet production deadlines or customers' delivery dates. He needs to be aware of the wider business picture, and this is particularly true when talking about maintenance management systems. These have not only to be good at the maintenance bits, but also have to be able to provide a wide range of management and financial information, which is increasingly being required to support the development of teamwork and empowerment. In addition, these systems are required to interface – or to be integral – with the company's ledger, its real-time process control systems, imaging and document management, payroll accounting, attendance monitoring, weighing and measuring, laboratory testing, the list goes on. The techniques for achieving this are mature and the appropriate integration of this data can add considerable value to the business.

One of the principal themes of this chapter will be that the implementation of systems needs to take account of business processes and the workplace culture, if the often substantial investment is to be realized. Competitive advantage comes less from the possession of this technology, than from the way in which it connects people together, both inside and outside the company, to maximize the use of human and other resources.

The introduction of computerization in the management of the maintenance function occurs more often as part of the *replacement* of an existing system, rather than as part of the development of a green field site, the latter being by far the easier task, in that there are no data to be transferred, no re-training to carry out, no prejudices to overcome, and no 'downsizing' to deal with. System replacement can occur for one or more of a variety of reasons – changes in the business environment, the requirement to automate or to cut costs, takeovers, product changes, etc. Often, existing systems are incapable of supporting new business requirements.

It is as well to think of what the expected benefits will be; auditors are increasingly taking an interest in IT expenditure. Pressure to reduce costs is often a driving force behind the introduction of technology, the intended end-products being fewer people – 'Downsizing' or 'Decruitment' (sic) – more automation, increased throughput, higher production, fewer rejects, etc. All of these benefits can be quantified, and should form part of the business case. Such changes in business culture may also be brought about by privatization, devolution, empowerment, teamworking, TQM, and so forth.

In many companies, the changes in technology discussed in the previous chapter have brought about the demise of the mainframe computer culture, which in turn has put the ability to introduce systems into the hands of users rather than IT specialists. This is both a good and a bad thing. Good because it is easier, quicker and cheaper; bad because it puts the often inexperienced user at the mercy of salesmen's promises – and a bad system can seriously damage or even destroy a business. The software industry is littered with systems which rose rapidly to fame and then died, systems which promised much but delivered little. IT projects, particularly the larger ones, do have a tendency towards failure rather than success.

Whatever the reader is involved in, somebody has been there before and has made the same mistakes. The risk of failure can certainly be minimized, however, and the ideas presented in the following pages should help the reader to do that.

There is, of course, nothing more potentially 'political' within a large company than IT, especially if the company has a central IT function. It is therefore as well to answer a few basic questions before embarking on what might otherwise be a perilous journey. For instance, is there a clear business strategy, elements of which the proposed system could help to meet? In other words, will replacing the maintenance management system help to meet any of the company's strategic aims? If it does, then the next step is to find out whether there is a clear IT strategy? There is no use spending large amounts of time looking at systems if some of them will not fit the technical architecture of the company.

It is worth remembering that maintenance management does not operate within a vacuum the best systems are highly integrated with other parts of the business and will be accessed widely for the information they

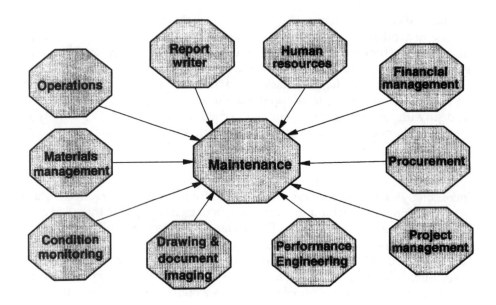

Figure 14.1 An integrated maintenance information system

contain. Figure 14.1 outlines a highly integrated* maintenance information management system (MIMS) which has a wide range of associated modules. Such a system is applicable across all types of process and heavy engineering industries – including mining, public utilities, petro-chemicals and steel production, as well as the manufacturing sector (from electronics to automobiles)

Selecting, implementing and commissioning a computerized maintenance system

One of the biggest mistakes that is commonly made is to dive straight into looking at what is on the market. The danger with this is that the potential buyer may be unduly influenced by the first system seen, before he has been able to stand back and assess the extent to which it meets the requirement.

It is important to realize that one of the main failings of system implementation projects results from choosing a system which the user-community feel no commitment to. As stressed earlier, large IT projects have a tendency to fail, so it is vital to attract and sustain the involvement of users at all levels in the decision-making process. This has the effect of

---

*An *integrated* system contains several different functions, e.g. maintenance, stores control, etc., and if one of these is changed this is immediately reflected, where relevant, in all the others.

tapping into the natural enthusiasm of operators and engineers from the very earliest stages and achieves a high level of commitment to the chosen systems (some suggestions for this will be discussed later).

Generally speaking, you should be seeking a packaged solution. The benefits of this are that the system is well tested, has had other users who can be talked to, and will be capable of being implemented in a reasonable time scale. The downside of buying a best-of-breed package is, of course, that it will only give you what your competitors either already have, or can themselves buy. You will need to look at these packages to see whether there is something special you can add, or make use of, which gives an edge. Bespoke software should be the last resort. In particular, do not contemplate an in-house development, re-inventing the wheel was never a good idea!

There are very many decisions involved in selecting a computerized maintenance system which will work for your business, and equally as many ways in which the enterprise can fail. Such a complex process benefits from a systematic approach, and needs to be related to clear objectives which can be specified as project deliverables and milestones.

Step 1: Formulating a project specification

The specification is an extremely important document. It is the basis on which potential suppliers will construct their bid, and eventually it will be the basis of the contract. It is also the basis on which the project will be judged in terms of meeting its objectives, and so it pays to be specific in your expectations. Vague statements of expected outcome will be used by contractors to minimize deliverables and maximize additional work and costs. Time taken at this stage will repay itself later on when difficulties or disputes arise (which they will!).

Broadly speaking, the specification falls into six sub-headings:

1. General considerations
2. System management functions
3. Functional requirements
4. Technical and user manuals, and other documents
5. Support
6. Benchmarks.

1. General considerations

Start with a general section covering the environment in which the system is to work and the work areas it is to cover. This is important in case the potential supplier has no direct experience in the industry. It may be appropriate to include a description of the current system – whether it be computer or paper- based, whether the industry involved is specialized or has unique features.

Any general rules surrounding the tender should be clearly stated. It may be that you require the potential supplier to visit one of your sites, or adhere to a particular set of quality standards. You may require (and certainly should ask for) a list of reference sites comparable to your own, or instructions relating to costing and what should be included (e.g. commissioning, training). You should certainly indicate what the tenderer should do if some of the specification cannot be met. You may reasonably ask that tenders should cover hardware, but have this part shown separately and make it clear that it may be sourced separately.

Depending on how much technical advice is available within the company it is worth asking for information regarding the standards and methods used in the development of the software, and any company standards which have to be adhered to. Some companies have preferred database suppliers like Oracle. If the system is to operate over a local or wide area network, then set benchmark standards for transferring data and formulate a general guide to response times (e.g. 95% of all transactions should take less than two seconds). Remember, if the system is required to handle sound, video or still images, video-conferencing or drawing files, these can cause considerable loading on the network. Carry out comparative benchmark tests of these functions if it is likely that network capacity is an issue. You will require some technical assistance from people who are able to set up a part of the network and monitor traffic using a network analyser.

The supplier must be given an idea of how many users the system is likely to be required to handle. One of the biggest mistakes you can make is to buy software that is simply not man enough for the job. A modern system can look very appealing but you need the reassurance that it already operates somewhere with the number of *concurrent users* (the maximum number who might be simultaneously using the system) you will require. Again, don't underestimate, build in plenty of scope for expansion to other parts of the business, or to other communities of users – e.g. the accounts, safety or sales sections. When you have decided the number of concurrent users you will require, get potential suppliers to show you an application of that size that is already in operation. Do not be taken in by an assurance that if it works on half the number, it will simply scale up. It probably won't!

A good system will soon become popular and attract all sorts of new users simply because it provides information which was not previously available. Accountants and senior managers will soon make use of management information if they know it is readily available. This will impact on the number of concurrent users on the system, as well as on the capacity of the hardware. A good system may take you into new areas because of the information that it will provide. You will begin to accumulate data on plant reliability as well as cost, which will facilitate cost–benefit analyses aimed at determining the optimum replacement of plant and equipment.

Don't underestimate the true cost of this sort of project. The initial software and hardware costs are only the tip of the iceberg. As a rule of thumb, you should estimate to spend at least twice the software cost on consultancy, writing software modifications, and on training and general hand-holding. In addition, there are your own staff costs. The project team will need to be full time for probably a year and will comprise something like five per cent of the user community. Do not be tempted to use lame ducks, the walking wounded, or people shortly to retire (although this last group will be very useful later on for data clean-up). The project team members are going to be a valuable asset to you once the system goes live, and should be the best you can get hold of and keep.

If you are replacing existing hardware, it is as well to check with the accountant whether there will be any depreciation still left on the books. This will have to be included in the project costing, as this will have to be written off. You may be able to get some money back on the old hardware, but it is unlikely to be more than two or three per cent of the original cost, if that. One other issue regarding cost, you should be prepared for the almost inevitable outcome that your chosen software will *not* be that of the *lowest* tender; it may in fact be that of the *highest* one. This can cause considerable problems for organizations that are locked in to a 'lowest tender' culture. It will be your final choice. So be prepared!

Getting management to sign on to the project can be difficult. The already mentioned early need to set up an implementation team will get line managers shaking their heads and muttering about staff shortages and budget cuts. It's very simple. If there isn't one hundred per cent management commitment to the project, don't go ahead – because all experience indicates that it will fail!

There is, of course, a hardware element involved in the overall performance of the system, and it is important to distinguish between that and the software capability. In order to minimize the overall cost of deciding in their favour, potential suppliers will sometimes underestimate the impact of running their software offerings on your existing hardware (i.e. existing file servers). As far as software is concerned, failure to demonstrate it in an environment similar to yours would be a major negative factor when the various offerings are being assessed. An important element is likely to be the database engine. Your company may already have a preference for Oracle, or DB2, or whatever. This should be stated, although it is wise not to rule out alternatives (as long as they are not too proprietary) in case better performance can be achieved. There can also be significant licence cost differences which should be taken into account.

One last point. The software market is littered with companies who rise quickly, burn brightly, and then disappear. It is worth enquiring whether a tenderer has lodged copies of his latest code with a third party (usually a lawyer) as a precaution in case his company should fail. An *escrow* (the legal term) arrangement of this kind can provide some comfort. And it goes without saying that you should enquire into the company's financial stability (via Dunn and Bradstreet or one of the other such services).

Hardware is usually less of an issue unless the system runs on something unusual or runs only on one platform. These days most good systems are fairly 'open' and will run on all major platforms and operating systems. It is as well also to give consideration to contingency and disaster recovery issues here. *Contingency* refers to what happens if a component (e.g. a memory module or a disk drive) were to fail. Some hardware platforms, notably DEC, provide a level of built-in redundancy. Clustering of machines and volume-shadowing of the data will mean that the failure of a processor or disk drive will not have any noticeable effect on the user. Failure in simpler hardware configurations can result in a system crash which will affect all users. Failure in the power supply should be similarly allowed for, and should cause no loss of – or damage to – data.

The hardware configuration may be an issue. If you are intending to operate over more than one geographical area, using a wide area network, the environment which you already have may dictate choices. The main advantage of a client–server architecture, for example (see Figure 13.2 et seq.), is that when a user on the network requests data to be sent from the file server to his workstation (i.e. his PC) only the data which is actually to be worked upon is sent rather than the whole database. This substantially reduces the risk of data corruption, and reduces the amount of traffic which would otherwise be present on the network.

2. System management functions

This part of the specification offers an opportunity to ensure that whatever software is offered has certain basic facilities. As was mentioned earlier, for reasons of flexibility of choice the system should be able to operate across several different types of hardware; also, it should operate within a recognized multi-user (network) environment, and should possess good security features at the 'front end' – i.e. protection by password of at least six characters in length, with the ability to force a change every month or so. In addition, it should be possible to restrict each user's access to only those parts of the system that are necessary for his particular purposes.

The system should contain a demonstrable data-archiving facility, and should be capable of carrying out on-line back-ups of data without interrupting the system use. It is also essential that data should be easily transferable to and from proprietary external packages like Lotus 123, Approach, DBase, Excel, etc.

3. Functional requirements

This is the real meat of the specification, where it is stated what the system is required to do. The general form of maintenance management systems was shown in the paradigmatic model outlined in Figure 12.1, a model which could be customized so that it would map the specific requirements of any particular industry. A nuclear power plant, for example, will have safety requirements relating to the monitoring of radiation exposure; chemical processing plants will have comparable, but different, needs. In these cases, maintenance systems that take little or no account of safety would be ruled out. Another requirement clearly demonstrable via Figure 12.1 would be for a stores system that was closely integrated with the identification, planning and execution of maintenance work.

It is probably as well to distinguish between 'must haves' and 'nice to haves', and beware of old ways of doing business being enshrined in the new system; this is an ideal opportunity to state what you want, not what you currently do. Remember, you are beginning a process which will affect the business for the next five to fifteen years, so try to anticipate likely changes. This is a golden opportunity to do a little business re-engineering. During the process of evaluating different tenders many new ideas will present themselves. Be open to them, remember world class businesses are constantly looking to improve the way they do things.

There is no standard pattern of describing what you require from the new system. It is probably enough to give a simple verbal description of each function, especially where the process is fairly generic. Stores systems, for instance, are much the same in most industries, although you may highlight certain features which are important to you (e.g. should engineers be able to 'bond' spares when they formulate a job so that those spares will be waiting when the tradesman goes for them?). Every industry seems to have its own special way of controlling maintenance work. If yours is one in which safety procedures are especially important – e.g. in which equipment needs to be locked off or tagged out and where permits-to-work are required – then there may be particular *mandatory* facilities within the system, e.g. if a job is altered after it is safety-assessed then it must be immediately routed back for re-assessment.

Interfaces[*] with other systems – either existing or planned – will add considerable value to the new system. It is as well to specify these in as much detail as possible, particularly if the system to be interfaced with is old or runs on a mainframe. Production of an interface with a new system is likely to be relatively easy; it will be little different from that for exporting and receiving data from a spreadsheet. Conversely, producing a mainframe interface program to feed, say, an old ledger system is likely to be very expensive. As already discussed, there may be many other interfaces (see Figure 14.1 *et seq.*). The main point to remember is that good packages will contain many or all of these features fully integrated with each other. The great value of integrated packages is that they do away with the large amounts of manual effort normally involved in transferring information from one system to another. Figure 14.2 shows a partially integrated system which also has interfaces with external systems.

Most modern maintenance management packages will have facilities for writing

---

[*]*Interfaced* systems are linked together, so that changes to one are transferred to the others. This does not happen immediately, and sometimes has to wait until a quiet period occurs because of the time needed for up-dating all the relevant files.

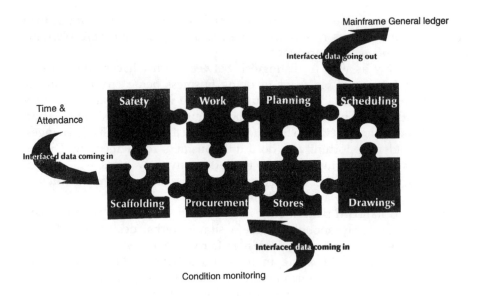

Figure 14.2 A partially integrated system

reports and creating graphs and charts (which can be imported into the text). Potential users of such facilities will need a good deal of training in the use of the report-writing programming language which is often employed. It is worth specifying a number of standard reports the ease of producing which can be assessed. All systems should likewise be able to export and import data to and from spreadsheets and external database systems – again, to facilitate comparisons, it is worth specifying such an activity, e.g. importing the number of maintenance jobs outstanding for more than a week, or exporting a list of routine maintenance tasks due in the next two weeks.

4. Technical and user manuals, and other documentation

It is not unusual to find very good software having very poor manuals (where the term 'manual' is used loosely, and covers the CD-ROM for example). Here the topic divides into two parts. First, test documentation will be needed by you and the supplier when carrying out acceptance testing of the delivered software – this, together with a test schedule, should be supplied by the vendor. Secondly, a specified number of user manuals will be required, together with technical manuals showing how the system is put together, delineating error routines, 'what to do's', etc. It is also a good idea to specify that the supplier provide quick-reference guides for users (one for each of them, plus spares). These are invaluable during the early implementation periods, and should be plasticized and pocketable. Subsequent software changes and bug fixes should result in updates to both technical and user documentation.

5. Support

Once the software is installed, and the initial warranty period (typically three months) has run out a software maintenance agreement will almost certainly be required. This can cost between ten and twenty per cent of the original software cost, and will guarantee provision of software updates, bug fixes, and a telephone help line (check that the hours of availability of this coincide with your work patterns). Get the tenderers to outline the extent of the help line service and determine whether it will be available out-of-hours. The existence of an established user group is a sign of a mature product, and is often a source of regular updates and improvements. You may find that the user group is industry specific, which can be very helpful.

6. Benchmarks

This is jargon for any standard of performance which it is crucial the system should meet. It is difficult to be specific as it will often refer to particular issues within your organization. For instance, some aspects of your process may be time-critical, so certain reports should be able to be produced on demand in no more than two minutes – or it should be possible to transmit large drawing files electronically. If there are particularly heavy requirements for report processing, you may wish to specify a maximum rate of slowing down of the response of other parts of the system while this is happening. Don't be tempted to compromise. You may decide that no slowing down is acceptable, and it is never a bad idea to make potential suppliers demonstrate the limits of their product by setting tough performance targets.

In the case of electronic transmission of drawing and other large data files there may be restrictions on the capacity of your local area network which require vendors to demonstrate methods of transmission that do not inhibit other traffic on the network. There are several techniques for compressing data, and these will vary from vendor to vendor.

You will gain valuable insight into the performance and usability of the system by visiting reference sites which are similar in size to your own. These should be visited once a shortlist of the possible vendors has been drawn up, and will provide an important source of information to help you reach a final decision.

Step 2. Setting up the tendering process

The first question which needs answering is 'Who shall we send the specification to?'. There are, of course, general software directories which list products by general function and will include sections on maintenance management. Specialist sources do exist, however, notably the *Maintenance Management Software Survey*, obtainable from Conference Communication (Monks Hill, Tilford, Farnham, Surrey, GU10 2AJ) in the UK. The Internet is another useful source and demonstration software can sometimes be obtained in this way. But be warned! – any software thus obtained should be vetted by a computer-virus checker before it is loaded on to any operational system.

It is not possible to make purchasing judgements on the basis of what is written in reviews, all you can do is issue your specification to as many possible suppliers as you think appropriate. This comes back to the importance of your specification. Checking the points in a detailed response against a good specification will make initial shortlisting a lot easier. If yours is an industry which has a co-ordinating body or organization which speaks for it, it is useful to enquire from it what systems are around. Do not necessarily be put off if a vendor does not have installations in your industrial sector. Similarities in maintenance requirements are often obscured by industry-specific language or jargon – one industry's annual overhaul is another's outage. There will be few industries in which the maintenance documentation system doesn't look very similar to the one that is modelled in Figure 12.1.

Generally speaking, there won't be that many vendors who are industry-specific. For instance, in the electricity generation sector there are probably only a handful of real contenders. The same is probably true for mining, petro-chemicals, steel making, automobile production, etc. In addition to package vendors, it is worth including one or two of the system integrators. These are companies, frequently large computer manufacturers or consultancies, who offer to meet your requirement by stitching together a number of packages to give you the best fit for your requirement. So if you already operate, say, a really good work management package, but rather poor stores-control or imaging facilities, the integrator will incorporate the best system in each category (as decided by either his judgement or your preference) and deliver a 'turnkey' solution – i.e. he does all the work, you just 'turn the key' and go.

The downside to system integration is likely to be the cost, but it is worth exploring. If you think all of this sounds rather daunting, you could employ an external consultant to do some or all of the legwork – which can be quite cost-effective compared with using your own time, provided he or she knows your industry. Depending on whether you use a small independent or one of the big consultancies, the cost of this can amount to between £500 and £1000 per day (1996 Western European charges). Again, try enquiring on the Internet. But remember, you must not give an external consultant any decision-making responsibility. The final decision is one that you and your staff will have to live with for some years.

Step 3. Tender evaluation and selection

Aim to invite no more than fifteen to twenty companies to tender. Although if you live in the European Union its internal legislation may force you to advertise in the *European Union Journal*, which will produce a large number of responses, many of them inappropriate.

Having invited your companies, ask them all to give a presentation to a selected audience of users. Involving the users may result in a conflict between what they want, and what may be best for the business in the longer term, a conflict which may not be easy to resolve, particularly within a company that does not have a consultative culture. But remember that success depends more on users' feeling involved than on buying the best system for your company. Ideally the two will be synonymous.

This approach will typically take at least half a day for each vendor. Find somewhere large enough to house as many of the users as are prepared to turn up. The only proviso is that they attend all of the presentations. Feeding them well is one way to ensure that they do! Provide them with a marking sheet which covers all of the main points in the specification, and don't let them leave any presentation until they have completed it! (See Appendix 2 for a model marking sheet.)

Don't forget to include any relevant specialists who may have a real stake in the outcome. The health and safety people in your organization may have issues which must be addressed. If there are non-users who wish to be involved, that's fine, but they don't get a vote! That includes your computer specialists, who may be influenced more by the technical sophistication of the software than by its practical utility.

Finally, hold the presentations over as short a period as possible, which helps users to make valid comparisons between the various offerings. It is worth setting aggressive time scales, because it focuses people's attention on the importance of what they are doing, and ensures that they are made available for the whole exercise.

Step 4. Selection

Provide a copy of the specification, together with the checklist itemizing the major headings in your specification, for your audience to use. At the end of the checklist ask whether the company should be shortlisted. When all the scores have been added up, you will probably get three or four clear contenders – often one clear leader plus a few also-rans (and a number of suppliers who can be ruled out very early on). The objective is to get general agreement on the shortlist before the last meeting breaks up – and make sure that its members see the final scores.

There is often a tendency, during the selection process, to favour companies who have experience in your sector. Try to avoid this. As already emphasized, maintenance issues are much the same wherever you go, so don't rule out a good system just because experience with it doesn't extend to your industry (although it may, of course, be ruled out for other reasons). Make a note of issues which come up during the presentations which were missed out from the specification, or which require additional emphasis. It is difficult to avoid users being dazzled by what a system looks like, rather than by its capabilities. Salesmen will naturally play up the technically sophisticated bits of their system because they look good. In this regard you will also need to be mindful of the level of computer literacy within your user community. Windows, multi-media, and mice may be inappropriate where the users are technically naive, are infrequent operators of the system, or work in a dirty environment.

The shortlisted companies should be invited back for a full-day presentation. Make sure the additional issues raised during the first round will be dealt with! Get the original audience back, or else you will spend much of the time covering old ground. While this is going on, reference sites for the shortlist should have been identified and visits arranged. Any problems arising from these visits should be fed back into the decision process. If a user group exists, go along to one of its meetings, or contact some of its 'movers and shakers'. Information thus derived will be invaluable in identifying what sort of product you are really dealing with.

Make sure you have identified the required functions that the chosen system does *not* have. These are the areas in which software modification will have to be considered if the additional functions are to be included. As a general rule, try to stick to the base product. This, almost certainly, will not be completely possible, but challenge every attempt to diverge from it; try to get people thinking about work-arounds, other ways of doing things which avoid changing the software. Remember, modifications are extremely expensive, and the more you have, the less standard the package becomes. There are also the hidden costs of modifications. When a new version of the software is released, your modifications will have to be re-worked, at a cost to you. It is definitely worth considering changing the way you do business, rather than getting involved in software modifications. Incidentally, make sure that the chosen vendor offers regular package updates. A good indicator is the existence of a strong user group.

Finally, do not allow the users to get drawn in to discussing the tendered prices. This is not their concern, and they should not be told the costs or allowed to be influenced by them. In the end, the actual cost of the software will be a minor factor in the overall cost of the project.

Step 5. Implementation issues

We have already talked about the implementation teams. For these a size of about five per cent of the user community is about right, so for a community of two hundred you will need a project team of at least ten, and this will certainly need to be boosted during intensive periods of data capture or training. Training the trainer is a good strategy for spreading the message; where you have more than one geographic location, use members of the original team to 'seed' further locations. Don't allow vital knowledge to be left with suppliers or consultants. The advantage of having a full time implementation team is that they become steeped in understanding of the system. This will significantly enhance their usefulness to the business and, therefore, their career prospects, as well as reduce future reliance on the supplier.

Naming the new system yourself can increase the sense of ownership in the user community. Hold a competition, publicize the results in your house newspaper if you have one, and make sure that the lead-in screen to the system is capable of carrying the new name.

Implementation will pass through a number of phases. Don't be tempted to leap in without a clear implementation plan; if you do you will certainly fail to meet deadlines. Appoint a project manager who should be as independent as possible and remember, he is your representative with both the supplier and the users. Consider using a good contractor, he can afford to be firm with people without worrying about his future prospects

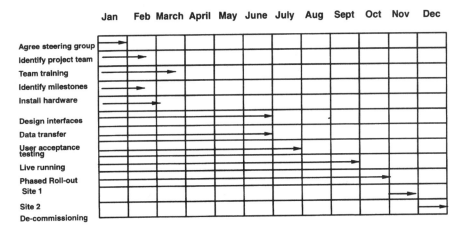

Figure 14.3   Implementation plan bar chart

within the company. His first tasks will be to produce the project plan (see Figure 14.3), allocate resources, and get this agreed by management.

Even in a relatively small organization, implementation can take some time. Eighteen months is not an unreasonable period given the need for data transfer and training, although it can be shorter. You will need to keep the business community posted. Keep the profile up with a newsletter on progress; this will help to maintain enthusiasm, particularly when things go wrong, which they probably will!

You will almost certainly already have a considerable mass of historical data which will be invaluable to start the new system with. Failure to exploit this will make the new system less usable, and will put off the day when it begins to make a real contribution to the business. Don't be tempted to say that 'we will build up the data as we use the system'. Don't delude yourself; this is an excuse for avoiding a difficult job.

When you examine your existing data you will find not only that it contains a lot of garbage but that it has gaps. You will need experienced personnel to tidy it up. Consider using staff who will shortly be retiring, or bring back retired employees. This is a job where experience counts and it needs to be started at the earliest point in the project. Some data can be transferred electronically, using a proprietary database as the intermediate stage to allow the data to be cleaned up. Some data may have to be captured by going out on the plant and recording it manually. This is especially important where the new system is organized around a relational database, because the more data it contains, the easier it will be to locate any item. In addition, there will be fewer gaps in the relationship between plant items.

Storekeepers are regularly confronted by a craftsman holding up some unidentifiable lump of metal and asking 'Have you got one of these?' (evidence suggests that forty per cent of stores counter enquiries fall into this category). A relational database can quickly search under plant type, manufacturer's reference number (which could be the only number the item has on it), plant location, etc. (see Figure 14.4). But if you don't put the data in, you can't retrieve it later! Capturing this information at the beginning will release the power of the new system and will certainly strike a chord with the users, who will have spent half their working lives queuing at the stores counter, or leafing through manufacturers' catalogues. Don't underestimate the time which data capture can take. As stressed earlier, it can be a major cause of project over-run, and you will almost certainly not get it right first time.

Step 6. Live running

During the implementation process many issues will have been raised and resolved. The consequence of this will often be observed, for example, in the revealed inappropriateness of existing written procedures (given that they do, in fact, exist!). These procedures may have been automated, changed, or even partially

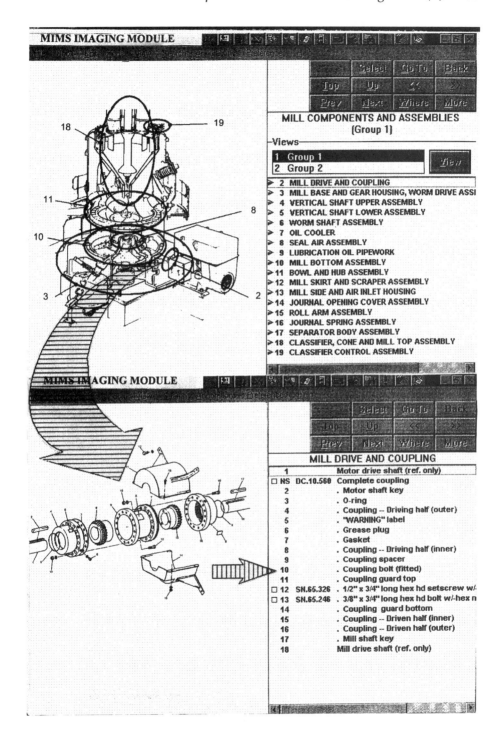

Figure 14.4   Exploded view taken from ImageLink, an MIMS module

eliminated by the new system. The maintenance documentation system modelled in Figure 12.1, for example, shows spares usage history being created from stores records. In a manual system this may involve record cards being filled in every day by the stores clerical staff. This entire step will be fully automatic within a computerized system, and reports will be produced at will. Many of these changes will not matter, but some will be critical, particularly if they are in any way safety-related. Set up a task within the project plan to produce new procedures and make sure it is done. Pay particular attention to new procedures covering computer systems administration, computer security and access control, and what to do in the event of a failure.

Once the system goes live, set up a user group and channel all change-control issues through it. If you have several sites, don't allow their various practices to diverge, there is too much at stake. Use peer pressure through your user group to obtain commonality. Don't fall for the 'but we are different' argument – they aren't, and they will have to change the way they behave.

A good system will quickly become an essential one, and an integral part of the production and management process. Once old paperwork is replaced there will be no immediate fall-back position in the event of a major failure of the system (or even a minor one). Don't leave decisions on contingency and disaster recovery until later. Contingency comprises two aspects. The first is the failure of a disk or processor – which can be dealt with, as already explained, by volume shadowing and clustering. Some hardware platforms may not offer this, or may have different solutions to the problem; it is important to know this up front. The second is a major hardware failure which puts the whole installation at risk. Given volume shadowing of data and clustering, this is unlikely unless you experience an actual disaster, i.e. fire, flooding, structural damage, etc. With proper off-site backup procedures in place the worst case should be the total loss of one day's data.

Disaster recovery can mean the difference between a business carrying on trading and its total demise. The loss of vital maintenance records or customer files can be crippling. It is always tempting to provide this sort of cover by having your own staff on emergency call-out, etc. – but don't do it! Go to the professional companies who make their living providing these services. They will offer documentation, and several trial runs a year, for less than it would cost to set up dedicated staff. They can provide a purpose-designed trailer, complete with network professionals, to bring the whole system back up within, say, twenty-four hours. It will be money well spent. Ask any of the City of London financial institutions who were bombed out of their business sites!

Finally, decommission the old system. Make sure that any residual historical data is properly archived and in a form which can be loaded into the new system if required. If the original hardware installation is a large one you may need to specify, prior to its removal, a disconnection procedure, particularly if you will have part-exchanged the old equipment for the new and the supplier will be wanting to get hold of it.

An example of a computerized maintenance management documentation system

At its simplest a computerized work management system is a repository for all the information about the physical plant – a record of all that has been done to it and details of the frequency and type of work that needs to be carried out. In other words, it is expected to carry out those functions indicated in Figure 12.1.

System outline

The day-to-day use of the system for managing maintenance work (see Figure 14.5) will consist of the registering and assessment of defects, the preparation of instructions for carrying out the work, and the associated safety and other special instructions. During this part of the process there is likely to be a considerable amount of searching for information. Sometimes it is not clear what part of the plant is involved, whether there is a 'standard' solution contained in the system, what spares are required, whether they are in stock, etc. A good system will save considerable time by allowing a search to be carried out 'sideways' to the main screen, i.e. by flipping out to, and back from, other parts of the system, gathering information as it proceeds. Once the engineering, stores, and work instructions have been completed, there are likely to be safety requirements which need to be met, either for special equipment or regarding isolation of the work area. Again, a good system will facilitate this, and will also hold standard isolation or *tag-out* routines (as they are sometimes called) to avoid their having to be re-specified.

Plant inventory

Most businesses maintain a register of their assets, and in an engineering environment this is often represented by some sort of hierarchical plant code. It is essential in any maintenance system that every job (and its costs) is linked to an asset. It is also important that when an asset is removed, either for repair or refurbishment, that it is possible to track its current location (e.g. in the workshop). When it comes back into service it is essential to know where it has been fitted. An electric motor may be removed from its position, replaced with another from stock, and sent away for repair. When it comes back, it may go back into store, or may be refitted elsewhere. Each asset, during its life, will acquire a history of work done on it and of the costs of this – which will be important information when deciding whether it is to be to repaired or replaced.

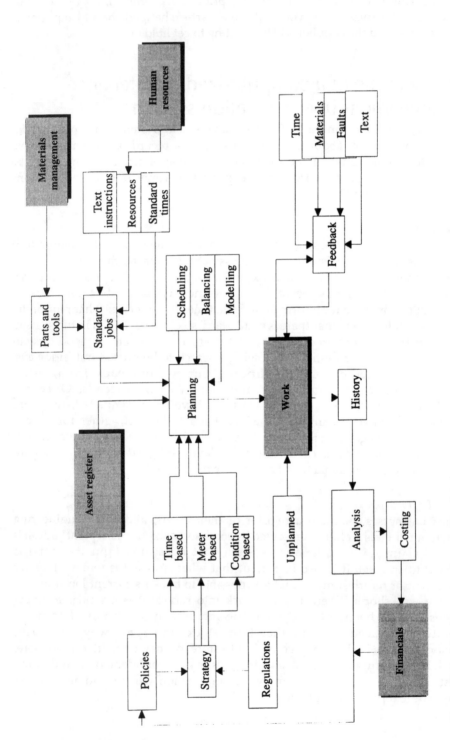

Figure 14.5   Maintenance work management system

Technical drawings and details

Recent advances in technology have meant that drawings, images, exploded diagrams, manufacturers' parts lists, etc. can be accessed as part of the database of information available to the maintenance practitioner from his maintenance management system. There is less need to be able to carry out drawing office functions on the system now that companies are receiving drawing information in electronic form from manufacturers. It is still useful, however, to be able to amend or 'redline' existing drawing records as plant modifications are made. Often, re-drawing is not done until a substantial amount of redlining has taken place. The result of all this is that the days of the company drawing office are coming to an end. Figure 14.6 outlines a document management system capable of handling and storing drawings, incoming mail items, manuals, files etc.

A system such as that of Figure 14.6 should, however, contain a method of controlling the version of any drawing in circulation, to prevent out-of-date issues being used. Paper copies of drawings and specifications tend to proliferate because they are so easy to duplicate. Accidents happen when work is carried out on plant which is not exactly as indicated on the drawing consulted. Some organizations require printed material to be stamped with a 'Use before....' date (as seen in supermarkets).

Inventory management

As already explained, there is an increasing need for stores management and spares control to be closely integrated into the basic maintenance

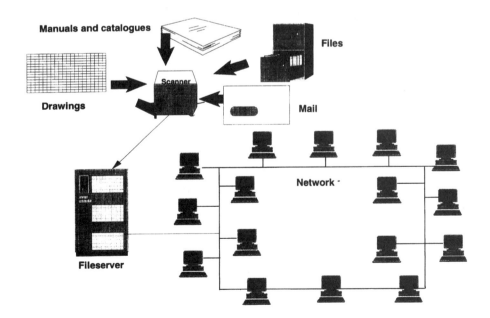

Figure 14.6   A document management system

function, thus eliminating much waiting and searching time. A good system will ensure that stores requirements are communicated electronically for making up *before* the craftsmen arrive. If the stores are large enough to warrant it, a picking list – perhaps with an integrated purchase order system – will be invaluable.

Ongoing work planning

When it comes to planning and carrying out work, it is very important to be able to know which staff are going to be available on site and – on the day – knowing which of those staff are actually available.

In Chapter 12 there is a general discussion of the various functions that constitute a paper-based work planning system – the role of the resource allocation board, for example. Within a computerized maintenance management system, these can be automated. Increasingly, industrial sites are installing electronic entry control systems. The information gathered from one of these can often be produced as an output file suitable for importing into the shift rota of the maintenance system. The further addition of information on holidays, training, etc. can facilitate the creation of a powerful tool for forward resource planning – thus facilitating the allocation of jobs to men, for instance.

Major shutdown planning

A major overhaul will certainly benefit by being properly planned and managed. Most good systems either have a project management tool, or provide easy interfaces to appropriate proprietary packages. Of course, it isn't just work that needs planning, but internal manpower and that of contractors. For this to be done effectively the system should contain a complete set of information on shift patterns – with holidays, training courses, etc. marked in. Such time and resource analysis is essential for the proper control and optimization of all projects, ensuring that they are completed on or before time and within budget.

As was pointed out in Chapter 12, overhauls – unless they are quite small – can generate many hundreds of jobs, many of them interrelated. The resulting complexities surpass the capabilities of simple wall-mounted bar-chart scheduling. A good maintenance management system will allow the production of trial schedules, so that time and resource constraints can be taken into account in the production of an optimal schedule. This is where historical information – especially from previous overhauls – is so valuable; to have accurate data on the time needed to carry out work, and on its cost, enables the maintenance engineer to avoid failing to complete the overhaul to time, thus avoiding lost production or cost overrun.

Additional features

A maintenance management system will provide the facility to track the life history of components, sub-assemblies, and whole plant units. This is a simple by-product of using the various functions of the system. Every piece of work has to be linked to a unit of plant, the information provided within the work order determining the level of detail subsequently available. This type of analysis enables complex decisions regarding further repair or replacement of major plant units to be resolved by analysing the lifetime costs. The record for an electric motor, for instance, would start with the date and price of its purchase, the date of its issue from stores, and its location on the plant. Repairs would then be recorded in terms of tradesman's time and cost of replacement components – until it was perhaps removed for refurbishment or rebuild. Its amended location (which might be the manufacturer's premises) would then be shown and, on its return, the date it was put back into stock, where it starts its cycle again. There will come a point where further repair is judged uneconomic. The method of finding that point may be to run a report every so often, which lists all components whose lifetime costs have exceeded a certain percentage of original purchase price. On even a small plant, such information can be very effective in keeping costs down. The same reporting facilities, given sufficient quality and completeness of data, can enable faults that recur with a particular type of item to be identified.

Safety monitoring

With the increasing pressure on companies world-wide to ensure that their staff operate within safe working conditions, a good system will have facilities for identifying work areas which require special consideration. The purpose of these is to create and maintain safety clearances in order to protect maintenance personnel while work is being performed. Clearances or *permits-to-work*, as they are sometimes called, are supplemented either by safety tags or padlocks, which are physically placed on the plant item. A few of the better systems will provide a facility for setting up, and using, standard clearances or isolations. More often than not it is small mistakes that cause large accidents, and one area in which a good system can be supportive is in controlling the safety assessment process. This should ensure that once a piece of work has been assessed, any subsequent change in the nature of the job description – perhaps by adding some small piece of extra work – will automatically remove all approvals, necessitating re-evaluation before they can be re-issued. It really depends upon the type of industry as to whether this is a serious problem or not.

Operational information and condition monitoring

Maintenance functions are making more and more use of real-time condition monitoring systems – in order to develop predictive strategies. The technology of these systems – which monitor vibration, temperature, and other characteristics of operational plant (see Figure 14.7) – is such that they can be put together by users rather than specialists.

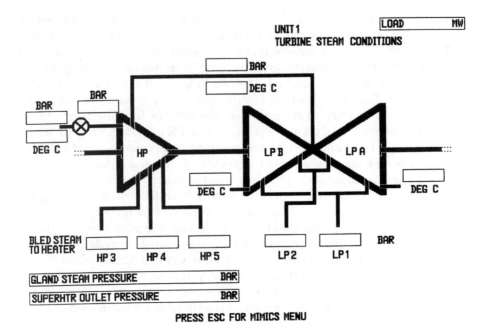

UNIT 1
TURBINE STEAM CONDITIONS

LOAD   MW

PRESS ESC FOR MIMICS MENU

Figure 14.7   Output from a condition monitoring system

Management information

Data only becomes information when it is organized in a form which is both meaningful and informative. Many systems come with an executive information tool (an EIS), but it may be anything from a simple report writing tool, to a sophisticated analytical capability that will produce graphs, tables of statistics, cost breakdowns, etc. However, there are also excellent tools available as components of systems such as Microsoft Office or Lotus SmartSuite. These provide spreadsheet or database software which can manipulate data extracted from most systems, and also integrate the results of this into word processing documents, thus providing a comprehensive reporting facility. There are benefits from an integrated system, but do not worry if it is not what you want. The answer can easily be obtained using one of these latter tools.

Figure 14.8 is a typical example of the type of graphic which could easily be inserted – along with pie charts, line diagrams and so on – into the text of a management report. It was created in a few minutes using the Freelance Graphics application software which is part of the Lotus SmartSuite package. There are several similar packages equally as good.

History

As already explained, a facility for building-up historical plant information (see Figure 14.9) will help engineers to make sensible judgements regarding

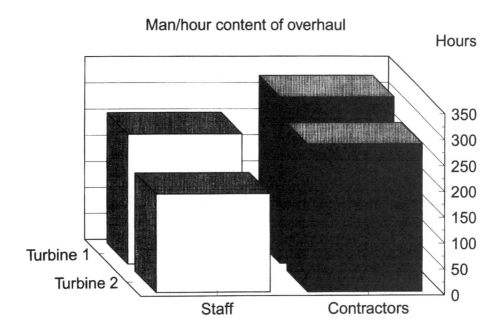

Figure 14.8 A simple graphic for a management report

whether to replace or refurbish plant – and that is why it is so important to capture as much data as possible as part of the implementation process, the facility then being immediately valuable the moment it goes live.

Recent developments in computer applications

Intelligent systems

Although they are still a long way from exhibiting true intelligence, there are several ways in which computer systems can imitate human decision-making, when they are then termed either *expert systems* or *artificial neural networks*, depending on the nature of their operation.

Both of these technologies are directed at providing some degree of 'intelligence' to any type of diagnostic situation which would otherwise have involved expert human judgement. Neither can be said to be truly intelligent, although their outputs can be very impressive. Expert systems operate on rules which are either developed as part of a dialogue with a human expert, or are based on previously documented cases. Because of the increasing amounts of data produced by real-time monitoring systems, and the decline (due to 'rationalization', 'downsizing', etc.) in the number of human experts available,

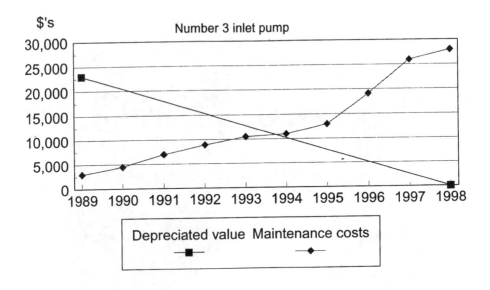

Figure 14.9   A display of cumulative maintenance costs

these systems offer considerable opportunities for extracting valuable new information from existing data. Most companies find themselves data rich, but information poor. Recent developments are able to take account of *qualitative* reasoning as well as of conventional *quantitative* modelling and diagnostics.

Figure 14.10 is a schematic of an expert system designed to undertake the water management functions of a hydro-electric scheme. The system rules – which indicate the actions to be taken during flood and other conditions to maximize power production but minimize the risks of flooding and other unacceptable outcomes – are based on many years of operator experience.

A further example is provided by an application of the COGSYS expert system software, one which is concerned with improving the operation and maintenance of the environmental plant within a stainless steel production facility. The plant produces around 300 000 tonnes of liquid steel annually, making stainless steel slabs for downstream rolling and finishing processes. The melting and refining processes often generate large quantities of airborne dust from the formation of metal oxides as the molten steel contacts air.

The aim of this application is to reduce the energy usage of the extraction system (see Figure 14.11), and the maintenance costs, by providing plant engineers with advice on extraction system performance and efficiency. Intelligent scheduling of filter-plant maintenance can provide an optimal strategy for changing filter bags, thus maintaining filter plant efficiency and extending bag life. In conjunction with a mathematical model the system will provide advice on the optimal use of extraction fans, dampers and filter compartments to reduce the overall cost of extraction and filtration. Finally,

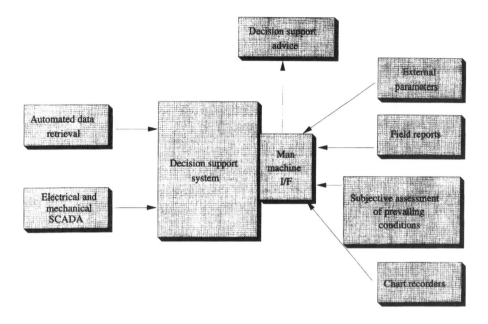

Figure 14.10   An expert system for water management

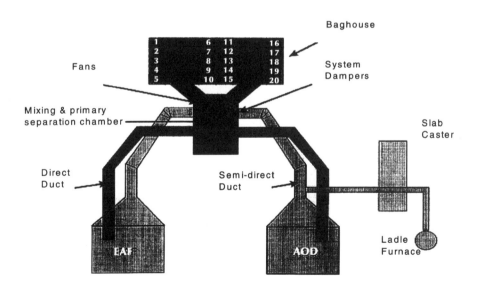

Figure 14.11   Steel plant fume extraction

by eliminating the possibility of fume build-ups in the melting shop, downtime will be reduced. Other applications in the plant are directed at steel making, quality prediction and blast furnace performance modelling.

As has already been said, the increasing degree of automation, and the reduction in the use of human operators, will increasingly require the use of this type of technology. However, capturing an expert's knowledge is one thing, replacing him with a computer and a semi-skilled operator is quite another. Expert systems require sets of rules, or cases for each situation, and although some current systems have the ability to infer and generalize – and in some instances to develop new rules by 'learning' – they are not currently good enough to completely replace human operators in decision-making situations, but do have a considerable and increasing role in decision support.

It is as well to be aware of these technologies, because software companies are increasingly offering additional modules of this kind. But do not underestimate the amount of work involved in building up the rules. Although some major plant manufacturers are getting increasingly interested in developing their own systems, using proprietary data, which may offer quicker entry into this technology, they will still have to decide whether they wish their plant knowledge and expertise to be in the hands of external suppliers. The constant underlying theme of this chapter has had to be:

> Use the best available standard software, but don't let the contractor walk away with any knowledge that is central to your business. You don't have to do it all yourself, but stay in control!

Appendix 1 Critical Path Analysis

Introduction

Critical Path Analysis (CPA) is the generic name given to graphical techniques for studying the interrelationships between activities in complex projects.

Some definitions

Activity

An operation consuming time only, or time and resources. An activity may be physical (such as the fabrication of an element of a structure) or it may be abstract (such as the accomplishment of a particular stage in a design calculation).

Event

A state of the project when all preceding activities have been completed and before any succeeding activity has started.

Graphical representation

Each activity in the project is represented by a directed line* which forms a link in the network. The direction and length of the line has no particular significance in representing the flow of time. Events are represented by circles at each end of an activity line. The arrowhead on the line identifies the events immediately preceding and following the activity (see Figure A.1). The identity of the activities is given by a label inside each circle, and so the line shown in Figure A.1 denotes the *ij* activity, linking the prior event *i* with the following event *j*.

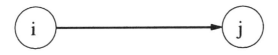

Figure A.1   Two events linked by an activity

---

* This method of representation, which has been adopted throughout this appendix, is known as *Activity on the arrow (AOA)*, as opposed to the alternative *Activity on the node (AON)* method. An explanation of the latter, with worked examples, was given in Kelly, A., *Maintenance Planning and Control* (pp. 232 et seq.), Butterworths 1984.

Interrelationship

The interrelationship between activities is embodied in the convention that an activity may not begin until all preceding activities in the same path are complete. Thus, in the example shown in Figure A.2 the event 3 is not regarded as having been reached until both the activities 1–3 and 2–3 have been completed, and until that stage has been reached the event 3–4 may not begin.

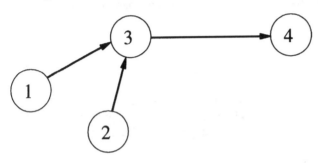

Figure A.2   • Inter-related activities

Start event

An event with succeeding but no preceding activities.

End event

An event with preceding but no succeeding activities.

Network

A graphical representation of the activities necessary for achieving the objectives of a project and showing their interrelationships. The nodes of the network are formed by those events which are common to two or more activities, and the terminals of the network are provided by the start event and the end event.

A useful convention in the construction of networks is to treat time as flowing from left to right, and to number the events so that for each activity the succeeding event has a higher number than the preceding event (see Figure A.3).

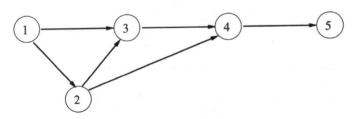

Figure A.3   A network

Dummy activity

In the example of Figure A.4, apart from the confusion of identity (both activities would be identified as 7–8) there would be other difficulties in the analysis of the network. These can be overcome by introducing a *dummy* activity into the network, as shown in Figure A.5. A dummy is a logical link, a constraint which represents no specific operation. It is represented by a dotted line to distinguish it from an activity and it carries an arrow to express the constraint. In Figure A.5 this facilitates the expression of the relationship as – *activity 9-0 may not begin until the parallel activities 7–8 and 7–9 have been completed* – while giving separate identities to the parallel activities.

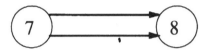

Figure A.4   Parallel activities with the same terminal events

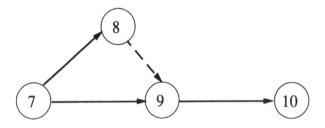

Figure A.5   Introducing a dummy activity

The example shown in Figure A.6 is a valid network construction only if the commencement of activity 8–10 and the commencement of 8–11 are dependent on the completion of 6–8 and the completion of 7–8. If the commencement of activity 8–10 depended only on completion of 6–8, and not on the completion of 7–8 an additional event, linked to event 8 by means of a dummy, would have to be introduced.

Example of network analysis

The first step in the practical application of network analysis is to reach agreement with other interested parties on the logic of the project to be analysed. It will be necessary to prepare a list of all the activities, and by careful enquiry to determine their interrelationships. It helps if the activities are listed in approximate chronological order. In a real project each activity would be identified by a short descriptive title, but for the purpose of this

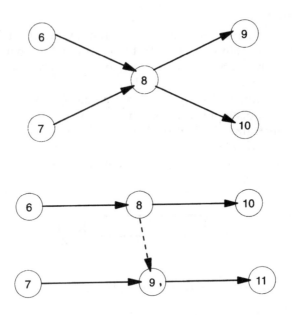

Figure   A.6

illustration the activities listed in Table A.1 are identified only alphabetically.

At this early stage it is often worth scrutinizing the interdependencies to see whether any redundant statements have crept in. For example, in Table A.1 we were originally told that activity I depends on the completion of activities A, C and H, but we were also told that activity H depends on the completion of activities A and C and so it is sufficient to state that the commencement of I depends on the completion of H. In the table, such redundancies have all been shown bracketed.

Table A.1   Network relationships

Activity	Interrelationships
A	Independent of all other activities
B	Depends on completion of A
C	Depends on completion of B
D	Depends on completion of B and C
E	Depends on completion of A
F	Depends on completion of B, C and E
G	Depends on completion of (C), D and F
H	Depends on completion of (A) and C
I	Depends on completion of (A), (C) and H
J	Depends on completion of B

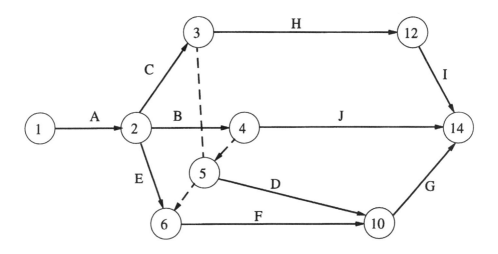

Figure A.7

The completed logic diagram is shown in Figure A.7. The most rigorous check that can be made of the logic consists of reading the constraints back from the diagram and comparing these with the original statements.

Activity times (duration)

Having constructed a network which embodies all the activities and interrelations between them our interest now turns to the time required for the individual activities and for the whole project. The time required to complete an activity is termed its *duration*. Estimated time or actual time may be used, depending on whether the network is used for forward planning or for maintaining progress on a project.

Critical path

Any uninterrupted sequence of events traced through a network constitutes a path. In any network there is always one path, from the start event to the finish event, which is of particular interest, the total duration of it being not less than that of any other path between the same two events. This is termed the *critical path*. The total duration of the critical path is the duration of the project, and the duration of the project is governed by the critical activities on this path.

To determine the critical path in a network it is necessary to examine the *earliest possible* and the *latest allowable* times for each event, after considering all possible routes through the network. The estimated durations for the example whose network has already been drawn are given in Table A.2.

Table A.2 Activity durations

Activity	Duration (days)
A	9
B	9
C	7
D	5
E	5
F	3
G	7
H	8
I	1
J	7

The earliest possible times are now calculated (see Figure A.8). Within each nodal circle the nodal ordinal number is given at the top, the earliest possible time at bottom-left, and the latest allowable time at bottom-right. The time of the start (node 1) is set to zero, and the times for other nodes determined by addition of the activity times. Thus

earliest time to node	2	=	0 + 9	=	9 days
earliest time to node	3	=	9 + 7	=	16 days
earliest time to node	4	=	9 + 9	=	18 days

There are two alternative routes to 5, i.e.

time to node 5 via node	3	=	16 + 0	=	16 days
time to node 5 via node	4	=	18 + 0	=	18 days

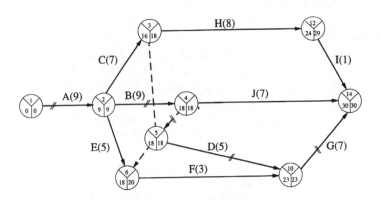

Figure A.8 *Completed representation of the network*

Clearly the route via node 4 determines the earliest time at which node 5 can be reached, if both paths have to be completed (which, from the network definitions, they must), so the earliest time for node 5 is taken as 18 days. After working through the whole network in this way it is found that the earliest time for the end event (node 14) is 30 days. We now know that the duration of the project will be 30 days, but the critical path has yet to be determined.

The latest allowable times for each event are found by taking the latest time for the end event (node 14) to be equal to the earliest time and then working back through the network to the other nodes, subtracting the activity times, thus

latest time to node	10	=	$30 - 7$	=	23 days,
latest time to node	6	=	$23 - 3$	=	20 days.

There are two routes connecting node 10 with node 5:

by direct route, latest time for node	5	=	$23 - 5$	=	18 days
by route via node 6 latest time for node	5	=	$20 - 0$	=	20 days

Clearly, node 5 must be reached by the eighteenth day if the project is not to be delayed.

Working through the whole network in this way the latest allowable times of all the events are determined. The latest time for the start event will be found to be zero. (If not, there is an error in the calculation.) An uninterrupted path can now be found through the network linking up nodes for which the earliest and latest times are equal. This is the critical path and the critical activities are marked on the network diagram by a double bar across each line. These are the activities which attract primary interest if the project time is to be controlled or reduced.

A measure of the importance of non-critical activities is given by the *float*, which is the time available for an activity in addition to its normal duration. The *total float* available to an activity is calculated from the earliest possible start time and the latest allowable finish time, and is the maximum additional time which can be absorbed by the activity without delaying the project. The *free float* available to an activity is calculated from the earliest start time and the earliest finish time, and is the additional time which can be absorbed by an activity without altering the floats available to other activities. These quantities are calculated via the following relationships:

total float   = (latest finish), (earliest start), (duration)
free float   = (earliest finish), (earliest start), (duration)

Thus, the total float for activity E is 6 days and the free float is 4 days. It can be seen that all the activities on the critical path have zero float.

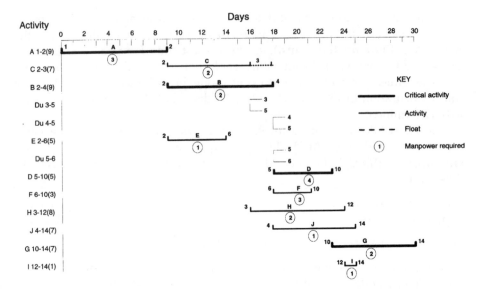

Figure A.9   Bar chart framework

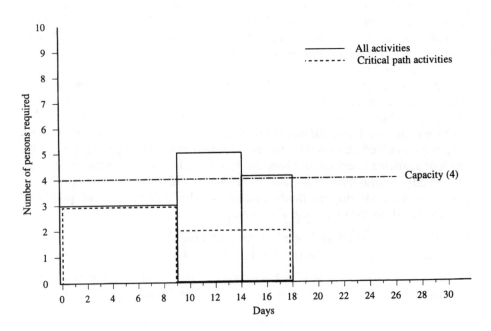

Figure A.10   Incomplete histogram of manpower requirement

Deriving a bar chart from the network

CPA has obvious advantages at the *planning* stage of a project. However, it lacks the bar chart's capacity

(i)   to represent time by the length of a line;
(ii) to be the vehicle for recording the progress of work (and hence controlling it).

Translation of the CPA network into a bar chart is therefore highly desirable. Essentially this is done by using the *head* and *tail numbers* of the activities to show the logical linkages between the activities, the head number being the nodal number at the *end* of an activity, the tail number that at the *start*. This procedure will now be illustrated, using the final network of the example, i.e. Figure A.8.

List the activities in order of increasing head numbers. Where two or more activities have the same head number, arrange these in order of increasing tail numbers. For each activity, list its manpower requirement, say. For our example this results in Table A.3.

Table A.3   List of activities by head and tail numbers

Activity Ref	(Tail no) – (Head no)	Duration (Days)	Manpower requirement
A	1–2	9	3
C	2–3	7	2
B	2–4	9	2
Du	3–5	–	–
Du	4–5	–	–
E	2–6	5	1
Du	5–6	–	–
D	5–10	5	4
F	6–10	3	3
H	3–12	8	2
J	4–14	7	1
G	10–14	7	2
I	12–14	1	4

Using the data in Table A.3 the bar chart, Figure A.9, and the manpower requirement histogram, Figure A.10, can be constructed.

Appendix 2 Specimen system-assessment score sheet

The following questionnaire should be used to assess each system by scoring the relevant questions on a scale of 1 to 10 (1 = poor, 10 = excellent). It is intended as an aid to shortlisting the systems viewed. A maximum of three systems will be selected and passed to a separate group for detailed scrutiny.

Name   _____

Location   _____

Job Title   _____

System Name _____

Date   _____ Session time (am/pm) _____

Functional requirements

Rate the following requirements for each module. The reference given in brackets relates to the specification document paragraphs if further detail is required.

Work control   **Score 1–10**

User authorization levels (Spec. section ref. no.)...

Repeated defect notification (Spec. section ref. no) ....

Restricted work order card modification (Spec. section ref. no) ....

Search/Sort on any field (Spec. section ref. no) ....

Job card grouping (Spec. section ref. no) ....

Bar code facilities (Spec. section ref. no) ....

Plant inventory asset register (Spec. section ref. no) ....

Routine work (Spec. section ref. no) ....

Statutory inspections (Spec. section ref. no) ....

Standard jobs (Spec. section ref. no) ....

Maintenance/ Operation instructions (Spec. section ref. no) ....

Outage/shutdown work control (Spec. section ref. no) ....

Engineering/Technical catalogue (Spec. section ref. no) ....

Spares list (Spec. section ref. no) ....

Recording of plant modification (Spec. section ref. no) ....

Safety Permit for work/tagout logging (Spec. section ref. no) ....

Defect analysis (Spec. section ref. no) ....

'What if' analysis (Spec. section ref. no) ....

Plant history analysis (Spec. section ref. no) ....

Materials management and procurement

Issues and returns (Spec. section ref. no.) ....

On-line enquiries (Spec. section ref. no) ....

Supplier's details (Spec. section ref. no) ....

Usage profiles (Spec. section ref. no) ....

Special orders (Spec. section ref. no) ....

Automatic re-ordering (Spec. section ref. no) ....

Audit trails (Spec. section ref. no)..

Work control link (Spec. section ref. no) ....

Personnel
Link to access control (Spec. section ref. no.) ....
Manpower levels (Spec. section ref. no) ....
Manpower cost against jobs/projects (Spec. section ref. no) ....
'On-site' indication (Spec. section ref. no) ....

Project management
Allocation of work to programme (Spec. section ref. no.) ....
Re-scheduling (Spec. section ref. no) ....
Multi-level programmes (Spec. section ref. no) ....
Critical path analysis (Spec. section ref. no) ....
Link to work control (Spec. section ref. no) ....

Electronic drawings/documents/catalogues
Retrieval by plant identifier (Spec. section ref. no.) ....
Sketching /red-lining facilities (Spec. section ref. no) ....
Scanning/printing size up to AO (Spec. section ref. no) ....
Workflow (Spec. section ref. no) ....
Version control (Spec. section ref. no) ....

Contract control
Contractor control monitoring (Spec. section ref. no) ....
Specific scaffolding/iInsulation contract control (Spec. section ref. no) ....

Budgets and finance
Estimates of expenditure for service levels (Spec. section ref. no.) ....
Financial statistics to profile assets (Spec. section ref. no) ....

Raw material management
Dynamic monitoring of usage (Spec. section ref. no.) ....
Manual over-ride facilities (Spec. section ref. no) ....
Reporting facilities (Spec. section ref. no) ....

Condition monitoring
Automatic job card from alarm condition (Spec. section ref. no.) ....
Reporting facilities (Spec. section ref. no) ....

Report generator
Interface to all modules (Spec. section ref. no.) ....
User specified reports (Spec. section ref. no) ....

General (Rate your overall impression of each module)
Work control ....
Stock control and procurement ....
Personnel ....
Project management ....
DM/Drawings/CAD ....
Contract control ....
Budgets and finance ....
Raw materials management ....
Condition monitoring ....
Report generator ....

Any other comments:

Should this package be short-listed to the top three?   Yes/No

Index